Lecture Notes in Educational Technology

Series Editors

Ronghuai Huang, Smart Learning Institute, Beijing Normal University, Beijing, China

Kinshuk, College of Information, University of North Texas, Denton, TX, USA

Mohamed Jemni, University of Tunis, Tunis, Tunisia

Nian-Shing Chen, National Yunlin University of Science and Technology, Douliu, Taiwan

J. Michael Spector, University of North Texas, Denton, TX, USA

The series Lecture Notes in Educational Technology (LNET), has established itself as a medium for the publication of new developments in the research and practice of educational policy, pedagogy, learning science, learning environment, learning resources etc. in information and knowledge age, – quickly, informally, and at a high level.

Abstracted/Indexed in:

Scopus, Web of Science Book Citation Index

More information about this series at http://www.springer.com/series/11777

Michail Giannakos

Editor

Non-Formal and Informal Science Learning in the ICT Era

 Springer

Editor
Michail Giannakos
Department of Computer Science
Norwegian University of Science
and Technology
Trondheim, Norway

ISSN 2196-4963   ISSN 2196-4971   (electronic)
Lecture Notes in Educational Technology
ISBN 978-981-15-6746-9   ISBN 978-981-15-6747-6   (eBook)
https://doi.org/10.1007/978-981-15-6747-6

This Springer imprint is published by the registered company Springer Nature Singapore Pte Ltd.
The registered company address is: 152 Beach Road, #21-01/04 Gateway East, Singapore 189721,
Singapore

Contents

Part I
Introduction to Science Learning

Chapter 1
An Introduction to Non-formal and Informal Science Learning in the ICT Era

Michail N. Giannakos

Abstract This chapter provides an overview of this edited volume on *Non-formal and Informal Science Learning in the ICT Era*. The goal of this volume is to introduce the reader to evidence-based non-formal and informal science learning considerations (including technological and pedagogical innovations) that have emerged in and empowered the information and communications technology (ICT) era. The contributions come from diverse countries and contexts (e.g., hackerspaces, museums, makerspaces, after-school activities) to support a wide range of educators, practitioners, and researchers (e.g., K-12 teachers, learning scientists, museum curators, librarians, parents, and hobbyists). The documented considerations, lessons learned, and concepts have been extracted using diverse methods, ranging from experience reports and conceptual methods to quantitative studies and field observation using qualitative methods. This volume attempts to support the preparation, setup, and implementation, but also evaluation of informal learning activities to enhance science education. In this first chapter, we introduce the reader to the volume, present the contributions, and conclude by highlighting the potential emerging technologies and practices connected with constructionism (e.g., the maker movement), coding, and joyful activities that are currently taking place under different spaces such as hackerspaces, makerspaces, TechShops, FabLabs, museums, libraries, and so on.

Keywords Informal learning · Non-formal learning · Science education

1.1 Introduction

According to the established definitions coming from the European guidelines (CEDEFOP 2009), formal learning occurs in an organized and structured environment (e.g., in an education or training institution or on the job) and is explicitly designated as learning (in terms of objectives, time, or resources). Formal learning is also intentional from the learner's point of view and typically leads to validation and

M. N. Giannakos (✉)
Norwegian University of Science and Technology (NTNU), Trondheim, Norway
e-mail: michailg@ntnu.no

© Springer Nature Singapore Pte Ltd. 2020   3
M. Giannakos (ed.), *Non-Formal and Informal Science Learning in the ICT Era*, Lecture Notes in Educational Technology,
https://doi.org/10.1007/978-981-15-6747-6_1

certification. This in the world of *science, technology, engineering, and mathematics (STEM) education* largely coincides with science classes in schools and tertiary education, although we agree that formal science learning plays an important direct and indirect role in non-formal and informal learning as well. The focus of this volume is on non-formal and informal science learning that takes place outside the classroom, and formal science learning is mentioned in cases where its contribution influences non-formal and informal science learning.

There is substantial broad knowledge already about informal science learning and science education outside the classroom (e.g., Lloyd et al. 2012; Falk et al. 2012; Robelen et al. 2011). What is still needed, especially at the European level, is much deeper insights into the nature and multifaceted impact of this type of learning. Gaining such deeply probing insights requires a focus on specific areas of the wider field, considering contemporary developments such as technological and pedagogical innovations, which will yield results that can then both be extrapolated and guide further research in other neighboring areas.

In non-formal science learning, we consider learning that is embedded in planned activities not always explicitly designated as learning (in terms of learning objectives, learning time, or learning support), but that contains an important learning element; non-formal science learning is, most of the time, intentional from the learner's point of view and can take place in museums, science camps/clubs, and so on. In informal science learning, learning results from daily activities related to work, family, or leisure, which is not organized or structured in terms of objectives, time, or learning support, and is mostly unintentional from the learner's perspective. Therefore, the level of intentionality plays an important role in both non-formal and informal science learning.

During the last few years, we have seen new ways in which non-formal and informal science learning is taking place through various activities (e.g., coding, making, play). Those activities are nowadays taking place outside K-12 school and higher education science classrooms, beyond the formal boundaries of science education. The increased interest in and implementation of those activities have led to the development and practice of different tools, affordances, and methods that support a wide range of educators and practitioners (e.g., K-12 teachers, museum curators, librarians, parents, and hobbyists). This chapter initiates a discussion on the role and potential of those activities to support non-formal and informal science learning, as well as on their impact on current practices and society.

1.2   Coding, Making, and Playing as Enablers of Out-of-Classroom Science Learning

Among the various informal science learning spaces and practices, much attention has been given to experiences and activities characteristically (one could also say, traditionally) associated with science museums and centers, zoos, exhibitions, competitions, field visits, and so on. However, the increasing emergence and proliferation of learning materials and practices emphasizing the joyful and creative element of informal science learning, as these are characteristically exemplified in coding, making, and joyful/play-based activities, have not yet drawn enough focus to them, while appearing to be one of the most important enablers in the field.

The links and contributions of coding- and making-based creative learning activities to science education are strong and intuitively obvious, albeit still only little explored and understood in depth. To a conservative approach to science education, coding and making may appear to lie beyond the boundaries of science classrooms, pertaining only to the fact that technology, engineering, and the arts are nowadays acknowledged partners of science and mathematics in the landscape of STEAM. However, the relation between these activities and science education, and especially informal science learning, is far deeper and very critical. Through computational thinking, design thinking, problem setting and solving, using their curiosity, imagination, creativity, critical thinking, and knowledge to understand and change the world, young coders and makers are at the same time deeply engaged science learners gaining insights into systems, data, and information, exploring patterns, getting involved in inquiry, collaborating and communicating, and understanding the role of science and technology in today's and tomorrow's societies and world.

1.2.1 Coding

Teaching coding to turn youngsters into confident and creative developers of digital solutions is currently gaining momentum in classrooms and informal learning spaces (coding fairs, labs, challenges, etc.) across the world. In 2013, the UK introduced a coding curriculum for all school students (Department for Education 2013); since then, several other European countries have been moving in the same direction. In particular, coding has, in recent years, become an integral part of school curricula in countries such as Estonia, Israel, Finland, and Korea. In the USA, a number of organizations (e.g., the acclaimed Code.org initiative) support computer programs in schools and offer coding lessons for everyone. Such new curricula and out-of-classroom initiatives are aiming far beyond just creating a new generation of computer programmers as a response to changing global demands for workplace skills. The purpose is to provide young people with the tools to navigate digital landscapes effectively, by developing their technological fluency and deeper understanding of how the digital world is created, how it might be used to meet our needs, and how we might repair or modify it. These growing efforts of governments to integrate coding as a new literacy and to support students in creative problem-solving tasks (Hubwieser et al. 2015) posit coding as a new and emerging affordance that has the potential to update and enable new non-formal and informal science learning practices.

1.2.2 Making

The maker movement of independent innovators, designers, and tinkerers has also dynamically entered the landscape of innovative education and informal learning (Papavlasopoulou et al. 2017). In makerspaces that are mushrooming in schools as well as in science centers, libraries, museums, and other informal learning spaces, more and more young makers are developing projects focused on prototyping innovations and repurposing objects. Maker education is emerging as a topical approach to interdisciplinary problem-based and project-based learning, entailing hands-on, often collaborative, learning experiences, and making in learning spaces and the positive social movement around it are seen as an unprecedented opportunity for educators to advance a progressive educational agenda. In the USA, the Obama administration strongly supported the growing maker movement as an integral part of STEM education, hoping to increase American students' ability to compete globally in the areas of science, engineering, and mathematics.

The confluence of the two movements, "coding" and "making," around the notion of digital making and fabrication is often linked to other technology-related learning activities such as those pertaining to robotics and the Internet of Things (IoT). Digital fabrication has dynamically entered the worlds of education and informal learning, boosted by worldwide FabLab initiatives (e.g., Stanford's FabLearn Labs, formerly FabLab@School). These educational digital spaces for invention, creation, inquiry, discovery, and sharing put cutting-edge technology for design and construction into the hands of young people so that they can "make almost anything," thus supporting project-based student-centered learning integrated into personal interests and daily life.

1.2.3 Playful/Joyful Activities

Across the spectrum of these emerging creative learning spaces, the elements of fun, joy, and playfulness are dominant. Especially outside classrooms, in the inviting and open-ended informal learning atmosphere of science centers, museums, libraries, zoos, community labs, outreach centers, fairs, contests, and so on, playful learning is the norm. There, fun and creative learning activities harness children's sense of joy, wonder, and natural curiosity, achieving high levels of engagement and learners' personal investment in learning. In a sense, in these informal learning spaces young people discover or reinvent their true selves as natural scientists, mathematicians, or artists, constantly seeking to construct new meaning and make sense of the world around them. Thus next to and far beyond game-based learning in science education (Li and Tsai 2013), whereby learning content and processes are incorporated into gameplay, in coding and making activities pure learning through play finds very fertile ground; as the seminal work by the LEGO Foundation (2017) puts it, "learning through play happens when the activity (1) is experienced as joyful, (2) helps children find meaning in what they are doing or learning, (3) involves active, engaged, minds-on thinking, (4) as well as iterative thinking (experimentation, hypothesis testing, etc.), and (5) social interaction." This is exactly what is happening when young people code and make in the context of playful informal science learning experiences.

1.3 Contributions and Themes of This Volume

1.3.1 The Lens of Science Capital to Understand Learner Engagement in Informal Makerspaces

Opportunities for young people to participate in making activities have increased dramatically in recent years. In describing informal learning spaces (e.g., science museums, makerspaces, FabLabs), many have argued that such spaces provide an inclusive approach to youth engagement in STEM education. The potential for enabling inclusive engagement is particularly significant given wider research findings that document the under-representation of some groups within the STEM workforce and engaged in STEM study, such as women and ethnic minority groups. Although the potential of making and makerspaces for empowering young people has been acknowledged, the ability of makerspaces to support equitable engagement is under-explored.

King and Rushton (this volume) draw on an underpinning framework that builds on the concept of science capital and the principles of the science capital teaching approach. In their contribution, they consider the ways in which makerspaces can be sites of equitable participation in informal science learning. They exemplify those ways through data from observations and interviews conducted in a UK-based makerspace, and argue that science capital pedagogic principles are evident in makerspaces and, when enacted, help to create an environment where young people feel valued and better able to participate in making and coding activities. King and Rushton (this volume) showcase how science capital pedagogical principles are utilized in makerspaces and argue that small changes to practice in the design and facilitation of makerspaces could result in such spaces being more equitable and socially just. Therefore, it is important for facilitators to empower children, as well as recognize and value the previous experiences children bring to the space and how these are incorporated into activities.

1.3.2 Digital Games as an Enabler for Science Learning

Digital games, online gamified labs, and virtual simulations (De Jong et al. 2014) present great potential for science learning, scientific literacy, and motivating interest in science. Such, mostly online, resources (e.g., https://www.golabz.eu; https://onl

inelabs.in) are most of the time free and enable children to experience science and math without having to set foot in an expensive, physical environment. There are resources in almost every science discipline that enable children to perform scientific experiments. Previous works have examined the effectiveness of such technological innovations in attaining learning objectives such as content knowledge, conceptual understanding, and problem-solving skills, usually in formal education settings.

Voulgari (this volume) examines the potential of digital games to support science learning and scientific literacy by looking at trends identified by previous meta-reviews over the past decade. Her work identified that there are games appropriate for most school subjects, including history and literature; however, research has focused on STEM-related games and learning objectives (e.g., physics, biology, chemistry, and the environment). During the last few years, there has been a shift to learning objectives and research that focus not only on content knowledge, but also on the understanding of scientific processes and practices, attitudes toward science, and higher order thinking skills. Factors involved in science learning through games have been identified such as the appropriate design of the game, individual characteristics such as previous science knowledge and interest, and the impact of the setting (e.g., a classroom environment).

1.3.3   Web-Based Science Learning: The Case of Computer Science MOOCs

Another opportunity that emerged during the last few years in science learning and non-formal learning is the rise of Massive Open Online Courses (MOOCs). MOOCs allow people to participate in a series of online learning materials, targeting specific content knowledge. There is research on the effect of MOOCs on learners' motivation, interest, and learning, as well as reasons for dropping out and disengaging. However, our knowledge about learners' preferences in the area of computer science and programming MOOCs is rather limited.

In their work, Krugel and Hubwieser (this volume) put into practice a MOOC in programming and investigate learners' experience by identifying aspects that improve or hinder that experience. In addition, they identify detailed reasons for dropping out of the MOOC in programming education. Overall, it is arguable that the design of the MOOC needs to be learner-centered and take into consideration the various particularities of the learners (e.g., timewise flexibility, interactive exercises). Such barriers seem to be of particular importance in the non-formal learning context, and further work needs to quantify their effect on learners' experience and adoption, as well as providing systematic ways of considering such aspects in the design phase.

1.3.4   Music and Coding as the Intersection of Literacies

Computational literacy has been defined by scholars such as diSessa (2018) and Vee (2017) and is currently gaining increasing attention and adoption in the science education field. This is also supported by the fact that computational tools and methods have become pervasive in modern scientific research across almost all fields of inquiry. What is less clear, however, is how to integrate computational literacy into formal, informal, and non-formal learning, as well as how to develop the next generation of computationally literate researchers.

Horn et al. (this volume) consider interviews, music, and computational artifacts produced by middle-school students in a summer camp setting using a learning platform called TunePad (https://tunepad.live). Their work furthers our understanding of the development of computational literacy through more informal learning experiences, with a focus on middle-school learners at the intersection of music and coding.

1.3.5   Non-formal Learning in Primary School: Programming Robotics

Besides the adoption of computational literacy in middle-school learners, during the last few years there has also been an ongoing and growing discussion about the necessity of such skills in primary education. The early development of key understanding, skills, and thinking approaches emerging from computational literacy and programming seems to have several positive effects on children. Programming plays a role in the context of formal, informal, and non-formal education, and more and more countries are including coding in their formal education (i.e., the curriculum), but also are developing various after-school activities and non-school organizations are developing concepts, methods, and activities. Despite the potential, it is still unclear to what extent and in what form computational literacy and programming can and should be introduced in primary education in the longer term, and the role that informal and non-formal learning activities can play in the transition and adoption period.

Geldreich and Hubwieser (this volume) investigate this further by conducting a series of interviews of Bavarian primary school teachers who put into practice programming activities with their entire class and in the non-formal setting of a programming club. Their work focuses on efficient practices and the challenges they encountered in these particular settings. A useful implication of their work is the view of teachers, who agree that all students should have the opportunity to learn programming—but that this has to be properly scaffolded and anchored to curriculum activities and learning materials.

1.3.6 Games for Artificial Intelligence and Machine Learning Literacy

Artificial Intelligence (AI) and Machine Learning (ML) education is also an interesting and rapidly developing field, attracting an increasing number of learners and instructors in the past few years. In response to this need, efforts in the USA, China, and other countries have resulted in AI/ML curricular activities for K-12 students (Touretzky et al. 2019). In addition, during the last few years, new online resources have been developed focusing on pre-college students, as well as professional development for teachers to learn the basics of AI (Touretzky et al. 2019). Recently, the Association for the Advancement of Artificial Intelligence (AAAI) and the Computer Science Teachers Association (CSTA) announced a joint initiative to develop national guidelines for supporting AI education in K-12 students. Moreover, initiatives such as the AI for K-12 working group (AI4K12) and AI4All (https://ai-4-all.org) were established to define what students should know and be able to do with AI, as well as to develop national guidelines and collect resources (e.g., videos, demos, software, and activity descriptions) for AI education in the USA.

General game playing is an exciting topic that is still young but on the verge of maturing, which touches upon a broad range of aspects of AI and ML. Giannakos et al. (this volume) conducted a literature review on the confluence of digital games and AI/ML education and created a general overview of games that have the capacity to support pre-college AI/ML education. The goal of this work is to provide a springboard for other scholars and practitioners to put into practice, experiment with, compare, and adapt the games and software listed to meet the needs of their students. The results depict how different games can enable opportunities for young people to engage with AI and ML, as well as for instructors and parents who want to teach a number of different concepts and topics in AI and ML.

1.3.7 Instructional Design of Non-formal Making-Based Coding Activities

Making has received growing interest in formal and non-formal science learning. However, the characteristics and design of such activities are not always clear or pedagogically efficient. Instructional models have been extensively used to align the design of learning activities with learning goals and objectives. Papavlasopoulou and Giannakos (this volume) illustrate and discuss the learning design of non-formal making-based coding activities, using the ADDIE instructional model. Utilizing the experience and results from empirical studies that have been implemented for over 3 years in the context of making-based coding workshops called Kodeløypa, they offer a set of best practices and lessons learned.

1.3.8 Games for Artificial Intelligence and Machine Learning Literacy

Access to technology and the ability to benefit from its use, as well as the skills and capabilities to innovate, design, program, make, and build digital technology, are all seen as pivotal for children's science learning. Makerspaces, FabLabs, and different kinds of coding clubs have started to offer children digital technology skills and competences. However, the potential of those environments in empowering children to make and shape digital technology remains poorly explored so far.

Kinnula et al. (this volume) investigate the potential of such environments to empower children to make and shape digital technology. The authors offer guidelines for practitioners working with children and their digital technology education in the context of non-formal learning and FabLabs. The special emphasis on these guidelines is enabling ways of working that respect and empower children. These guidelines should be useful for both teachers and facilitators when planning and implementing children's projects in FabLabs, with special emphasis on school visits to FabLab premises. The insights of this chapter should be useful broadly for researchers interested in the empowerment of children to make and shape digital technology through design and making, as well as for FabLab personnel—instructors and managers alike—and for teachers or city administrative staff who plan to work in collaboration with a local FabLab.

1.3.9 Conceptualizing Science Education and Its Ecosystem in Non-formal and Informal Settings

In the closing chapter of this volume, Giannakos proposes a conceptualization of informal and non-formal science education through an ecosystem model. The conceptualization of science learning in the form of an ecosystem is not new (Traphagen and Traill 2014; Corin et al. 2017), but it is arguable that it provides both the language to discuss an inclusive learner-centered system and the roadmap to develop collaborations between organizations and groups in the future (Corin et al. 2017). The learning ecosystem perspective aims to improve our current understanding of how various factors need to cooperate, coordinate, and collaborate to enable efficient and meaningful science learning in informal and non-formal learning settings.

1.4 Conclusions and the Way Ahead

The advances in technologies, manufacturing equipment, and learning spaces offer diverse opportunities for non-formal and informal science learning, especially when supported by coding, making and engaging, and joyful practices and designed in an appropriate pedagogical manner. From current research, it is difficult to tell what aspects of environments, technologies, applications, equipment, and practices can have a positive impact.

The current drive in many countries to teach STEM subjects to young people has the potential to further research initiatives into how information and communications technology (ICT), practices, and spaces have the capacity to enable non-formal and informal science learning. However, there are a number of challenges in ensuring that procedures/practices, tools, and environments embody appropriate progression and engender motivation and joy, which are critical for non-formal and informal learning contexts.

To explore the future of various spaces and ICT tools to foster engagement and creativity in science learning, we seek to promote interest in contemporary tools, practices, and affordances, such as computing and coding, and to put them into practice in different spaces such as hackerspaces, makerspaces, TechShops, FabLabs, and so on. This will allow us to better understand and improve their qualities as well as to accelerate the process of disciplinary convergence. In this volume, we present different works, coming from researchers with different backgrounds, showcasing the importance of disciplinary convergence. Bridging relevant disciplines such as learning sciences, science education, computer science, and design, among others, has the capacity to encourage ambitious research projects tackling the major themes of science education, including educational policy, instructor development, emerging science literacies, theory development, science learner empowerment, the development of appropriate environments and technologies, and practice development, to mention but a few.

Acknowledgements This work is supported by the "Learning science the fun and creative way: coding, making, and play as vehicles for informal science learning in the 21st century" Project, under the European Commission's Horizon 2020 SwafS-11-2017 Program (Project Number: 787476) and the "Learn to Machine Learn" (LearnML) project, under the Erasmus+ Strategic Partnership program (Project Number: 2019-1-MT01-KA201-051220).

References

CEDEFOP. (2009). *European guidelines for validating non-formal and informal learning.* Luxembourg: Office for Official Publications of the European Communities. Retrieved from https://www.cedefop.europa.eu/EN/publications/5059.aspx.

Corin, E. N., Jones, M. G., Andre, T., Childers, G. M., & Stevens, V. (2017). Science hobbyists: Active users of the science-learning ecosystem. *International Journal of Science Education, Part B, 7*(2), 161–180.

De Jong, T., Sotiriou, S., & Gillet, D. (2014). Innovations in STEM education: The Go-Lab federation of online labs. *Smart Learning Environments, 1*(1), 3.

Department for Education. (2013). *Statutory guidance: National curriculum in England: Computing programmes of study.* Retrieved March 1, 2020, from https://www.gov.uk/government/publicati

ons/national-curriculum-in-england-computing-programmes-of-study/national-curriculum-in-england-computing-programmes-of-study.

DiSessa, A. A. (2018). Computational literacy and "the big picture" concerning computers in mathematics education. *Mathematical Thinking and Learning, 20*(1), 3–31.

Falk, J., Osborne, J., Dierking, L., Dawson, E., Wenger, M., & Wong, B. (2012). *Analysing the UK science education community: The contribution of informal providers.* London: Wellcome Trust.

Hubwieser, P., Giannakos, M. N., Berges, M., Brinda, T., Diethelm, I., Magenheim, J., et al. (2015). A global snapshot of computer science education in K-12 schools. In *Proceedings of the 2015 ITiCSE on Working Group Reports* (pp. 65–83).

LEGO Foundation. (2017). *What we mean by: Learning through play.* Retrieved from https://www.legofoundation.com/it-it/who-we-are/learning-through-play.

Li, M.-C., & Tsai, C.-C. (2013). Game-based learning in science education: A review of relevant research. *Journal of Science Education and Technology, 22*(6), 877–898.

Lloyd, R., Neilson, R., King, S., Mark Dyball, M., & Kite, R. (2012). *Science beyond the classroom: Review of informal science learning.* London: Wellcome Trust.

Papavlasopoulou, S., Giannakos, M. N., & Jaccheri, L. (2017). Empirical studies on the maker movement, a promising approach to learning: A literature review. *Entertainment Computing, 18,* 57–78.

Robelen, E., Sparks, S., Cavanagh, S., Ash, K., Deily, M.-E., & Adams, C. (2011). Science learning outside the classroom. *Education Week, 30*(27), S1–S16.

Touretzky, D., Gardner-McCune, C., Breazeal, C., Martin, F., & Seehorn, D. (2019). A year in K-12 AI education. *AI Magazine, 40*(4), 88–90. https://doi.org/10.1609/aimag.v40i4.5289.

Traphagen, K., & Traill, S. (2014). *How cross-sector collaborations are advancing STEM learning.* Los Altos, CA: Noyce Foundation.

Vee, A. (2017). *Coding literacy: How computer programming is changing writing.* MIT Press.

Michail Giannakos is a Professor of interaction design and learning technologies at the Department of Computer Science of NTNU, and Head of the Learner-Computer Interaction lab (https://lci.idi.ntnu.no/). His research focuses on the design and study of emerging technologies in online and hybrid education settings, and their connections to student and instructor experiences and practices. Giannakos has co-authored more than 150 manuscripts published in peer-reviewed journals and conferences (including Computers & Education, Computers in Human Behavior, IEEE TLT, Behaviour & Information Technology, BJET, ACM TOCE, CSCL, Interact, C&C, IDC to mention few) and has served as an evaluator for the EC and the US-NSF. He has served/serves in various organization committees (e.g., general chair, associate chair), program committees as well as editor and guest editor on highly recognized journals (e.g., BJET, Computers in Human Behavior, IEEE TOE, IEEE TLT, ACM TOCE). He has worked at several research projects funded by diverse sources like the EC, Microsoft Research, The Research Council of Norway (RCN), US-NSF, German agency for international academic cooperation (DAAD) and Cheng Endowment; Giannakos is also a recipient of a Marie Curie/ERCIM fellowship, the Norwegian Young Research Talent award and he is one of the outstanding academic fellows of NTNU (2017–2021).

Chapter 2
Applying the Lens of Science Capital to Understand Learner Engagement in Informal Maker Spaces

Heather King and Elizabeth A. C. Rushton

Abstract Opportunities for young people to participate in making activities—either within school-based learning or within the growing number of makerspaces being established outside of formal education—have increased dramatically in recent years. Whilst some have advocated young people's participation in makerspaces as an opportunity to democratise access to STEM learning, it is also acknowledged that these spaces reproduce patterns of inequitable participation found in other science-related settings. An underpinning framework that builds on the concept of science capital and the principles of the science capital teaching approach may help a better understanding of this issue. Drawing on data from observations and interviews conducted in a UK-based makerspace, we argue that science capital pedagogic principles are evident in makerspaces and, when enacted, help to create an environment where young people feel valued and better able to participate in making and coding activities. We argue that small changes to practice in the design and facilitation of makerspaces could result in such spaces being more equitable and socially just.

Keywords Makerspaces · STEM · Science capital · Science capital teaching approach · Making · Coding · Facilitators · Equity

2.1 Introduction

Opportunities for young people to participate in making activities—either within school-based learning or within the growing number of makerspaces being established outside of formal education—have increased dramatically in recent years. In describing informal makerspaces, many (Brahms and Crowley 2016; Calabrese Barton et al. 2017; Honey and Kanter 2013; Martin 2015; Sheridan et al. 2014)

H. King (✉) · E. A. C. Rushton
School of Education, Communication and Society, King's College London, London, England
e-mail: heather.1.king@kcl.ac.uk

E. A. C. Rushton
e-mail: elizabeth.rushton@kcl.ac.uk

© Springer Nature Singapore Pte Ltd. 2020   15
M. Giannakos (ed.), *Non-Formal and Informal Science Learning in the ICT Era*, Lecture Notes in Educational Technology,
https://doi.org/10.1007/978-981-15-6747-6_2

have argued that such spaces provide an inclusive approach to youth engagement in STEM (science, technology, engineering and mathematics) education. The potential for enabling inclusive engagement is particularly significant given wider research findings which document the under-representation of some groups within the STEM workforce and engaged in STEM study, e.g. women (Archer et al. 2013) and ethnic minority groups (DeWitt et al. 2010).

Making activities are variously defined. Honey and Kanter (2013) highlight their hands-on nature and their collaborative iterative approach to learning. Blikstein (2013), meanwhile, references the role played by technologies and materials in making endeavours. Calabrese Barton et al. (2017) definition draws attention to elements of collaboration and creativity in the making process. Martin (2015) offers a working definition of making as:

> …a class of activities focused on designing, building, modifying and/or repurposing material objects, for playful or useful ends, oriented toward making a 'product' of some sort that can be used, interacted with or demonstrated. (p. 31)

People who participate in making often self-identify as 'makers': the places in which makers come together to engage in making thus become makerspaces. Sheridan et al. (2014) describe makerspaces as, 'informal sites for creative production in art, science and engineering where people of all ages blend digital and physical technologies to explore ideas, learn technical skills and create new products' (p. 505).

The potential of making and makerspaces for empowering young people has been acknowledged (Halverson and Sheridan 2014). Indeed, Blikstein (2013) argues that the makerspace movement offers opportunities to democratise the production of twenty-first-century technology. Others, however, have questioned the notion that makerspaces support equitable engagement. Martin et al. (2018), for example, have highlighted making's 'equity problem' (p. 36), arguing that whilst making and makerspaces are frequently described as open to all, in reality they are far from representative. Researchers have also noted that a deep and full understanding of the potential of makerspaces for STEM engagement is currently limited, as an established body of research and theory has yet to develop. In this chapter, we seek to contribute to emergent theorisations pertaining to makerspace practice and pedagogy. We argue that the theoretical lens of the science capital teaching approach offers a framework for examining pedagogical interactions within makerspaces and for determining the extent to which they engender equitable engagement in STEM.

We begin this paper by setting out the theoretical concept of science capital and the related science capital teaching approach.

2.2 The Lens of Science Capital

2.2.1 The Concept of Science Capital

The concept of science capital builds on the work of Bourdieu (1977, 1986) and his theorisations of cultural capital. Bourdieu identified three interacting concepts as the key components in one's cultural capital: their capital (defined as social, cultural, financial, and symbolic resources); their habitus (their highly ingrained dispositions or attitudes); and the field (the wider socio-spatial arena which incorporates all power relations, social rules, and regulations). The notion of science capital, developed, refined, and validated by Archer and colleagues (Archer et al. 2015a; DeWitt and Archer 2017) refers to one's science-related resources and dispositions, and how these are valued in the field of science. The concept draws attention to the variation in resources, attitudes, social contacts, and relationships possessed by a learner that in turn helps them to 'get on' in science, or not. Some students, for example, are able to utilise or exchange their science-related resources and dispositions in learning situations. Others do not have the particular knowledge, contacts or dispositions that are expected and valued in science settings: their resources do not fit. The concept of science capital thus helps to explain why some students feel comfortable in science learning settings and see themselves able to participate in science-related study or careers in the future, whilst others do not feel comfortable: they do not see science as something for them.

2.2.2 The Science Capital Teaching Approach

In addition to explaining varied participation, a science capital perspective can help to broaden our understanding of how learning and engagement may best be supported. It directs attention to the ways in which particular resources are valued over others. It highlights the role played by the wider field in determining what counts, or not, as scientific behaviours. It challenges educators to consider the ways in which learning settings are structured and the extent to which they favour learners from dominant social backgrounds (Archer et al. 2015b). It encourages educators to reflect on the norms and expectations of what constitutes engagement (Godec et al. 2018). Below, we summarise the key tenets of the science capital teaching approach (see Fig. 2.1)— a framework developed in partnership with teachers (Godec et al. 2017)—built on the principles of a science capital.

The science capital teaching approach is based on a foundation of broadening what counts. This involves reflecting on the wider participation structures of the class-room and, significantly, valuing the varied intellectual and social resources that are embedded in individual learners' everyday practices. By acknowledging the diversity of resources, educators concomitantly reflect on the expectations placed on learners

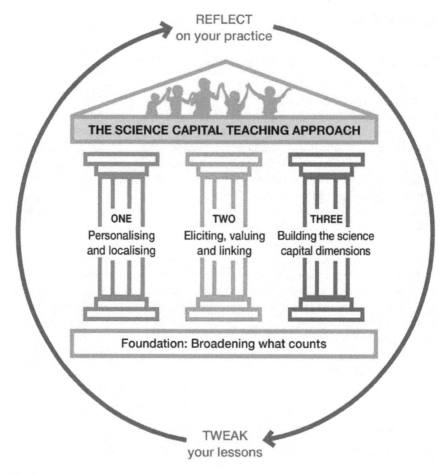

Fig. 2.1 The science capital teaching approach (Godec et al. 2017)

and the ways of performing that signify success: they thus broaden what counts as 'science' behaviour.

From a basis of broadening what counts the science capital teaching approach then has three pillars. These further seek to allow students to utilise their varied resources, feel valued, and thus able to engage. The pillar of personalising and localising highlights the necessity of making content personally relevant to a learner's life and way of knowing. It involves making explicit associations between a learner's experiences at home and in their community to aspects of the content being presented. Personalising and localising enables learners to see that the content can fall within their worldview, and that it can be of relevance to someone like them. The pillar of eliciting, valuing and linking highlights the particular practices educators must employ to support and value learners' knowledge and thereafter link this knowledge to specific content. The third pillar forming the science capital pedagogical practice refers to the need

1. **Scientific literacy**: a student's knowledge and understanding about science and how science works. This also includes their confidence in feeling that they know about science.

2. **Science-related attitudes, values and dispositions**: the extent to which a student sees science as relevant to their everyday life.

3. **Knowledge about the transferability of science**: understanding the utility and broad application of scientific skills, knowledge and qualifications.

4. **Science media consumption**: the extent to which a student engages with science-related media including, television, books, magazines and internet content.

5. **Participation in out-of-school science learning contexts**: how often a student participates in informal science learning contexts, such as at science museums, science clubs and fairs.

6. **Family science skills, knowledge and qualifications**: the extent to which a student's family have science-related skills, qualifications, jobs and interests.

7. **Knowing people in science-related roles**: the people a student knows (in a meaningful way) among their wider family, friends, peers and community circles who work in science related roles.

8. **Talking about science in everyday life**: how often a student talks about science with key people in their lives (e.g., friends, siblings, parents, neighbours, community members).

Fig. 2.2 The science capital dimensions (Archer et al. 2015a)

to incorporate science capital dimensions into the design of learning experiences. The dimensions (see Fig. 2.2) describe particular practices or dispositions which are have been found to correlate with an increased propensity for an individual to feel comfortable with science (Archer et al. 2015a). By peppering interactions and learning opportunities with comments about, for example, the value of science for society (dimension 2), or that skills and qualifications in science can open doors to many different sorts of jobs (dimension 3), such notions become normal, enacted, and over time will contribute to a growth in one's science capital.

2.2.3   Applying the Concept of Science Capital to Makerspaces

The science capital pedagogical approach was originally developed to support science teachers working in classroom settings. However, it has also been adopted by museum educators as a framework to support their facilitation of science and technology content (DeWitt et al. 2018). In the UK, national bodies promoting science and engineering have also applied the principles to their practice.

We suggest that the principles—broadening what counts, personalising and localising, eliciting, valuing and linking, and seeking to underscore particular practices, dispositions and understandings—can similarly be used to support makerspaces become more equitable sites of STEM engagement. Indeed, we suggest that adopting pedagogical practices based on the concept of science capital may be easier for an informal makerspace setting than a school. After all, classrooms are bound by historical and administrative structures. The emergent field of making is arguably less constrained and, potentially, has many more opportunities to design and re-figure learning interactions that are pro-active in ensuring equity.

In the sections below, we discuss makerspace practices through the lenses of science capital and the science capital teaching approach. In particular, we reflect on the ways in which extant pedagogical practices in makerspaces may be further enhanced to promote equitable engagement. Our data draws from in-depth observations of one inner city makerspace and interviews with makerspace educators. Our data collection was guided by the following research question:

-To what extent do makerspaces engender equitable access to the STEM learning opportunities inherent in the space?

2.3   Research Study

Here, we describe our research site, data collection methods and participants before outlining the analytical process used in this study.

2.3.1   The Research Site

Data for this study was gathered from a social enterprise located in south-east England. The initiative (developed over the last 5–10 years) aims to provide young people with greater opportunities to be imaginative and creative and views such practices as a vital part of both a child's development and a fundamental part of their learning. The makerspace facility offers young people aged 6–18 the chance to participate in a wide range of learning activities including coding and making. The activities include school workshops, family activities, holiday club sessions and community events. Sessions range from a drop-in, where families may spend a few hours, to half-day or whole-day school visits or holiday clubs. The activities are predominantly provided free of charge to participants. The sessions for which participants are required to pay are either very low cost (e.g. family drop-in sessions) or include free places for those from low-income groups (e.g. holiday club sessions). The programme of activities additionally includes sessions that are specifically aimed at young people with special educational needs, as well as those children who are home educated.

2.3.2 Data Collection

Our data derives from three main sources: (1) observations and field notes; (2) informal discussions with young people and teachers; and (3) semi-structured interviews with six key informants.

Prior to the collection of data, observation and interview schedules were developed and agreed by partners from across our EU-funded project, COMnPLAY-Science.[1] Our observations took place over 35 hours between April–August 2019 and involved observing a range of one-off facilitated coding and making activities. During each activity, we remained on the edge of the learning space and took field notes by hand. On occasion, we engaged with young people, their teachers and parents and the facilitators. This took place at the prompting of participants, e.g. when we were asked a question by a child or parent, or when we were invited by a facilitator to take part in an activity. At the regular points during observations, we discussed our notes and completed initial reflections.

We conducted six semi-structured interviews with key informants. Three were female, three were male and all were aged between 20 and 45 years of age. These key informants were recruited to gather a detailed understanding of making and coding in informal spaces and were identified through our networks. As Braun et al. (2009) have described, key informants typically provide 'in-depth experience and knowledge-based perspectives on under-researched topics' (pp. 113–114). The key informants in our research occupied positions 'inside' the making and coding communities and as such, were members of the communities of practice about which they were speaking. This contrasted with our position as researchers seeking to understand the pedagogy of making and the practices of makerspaces. The accounts shared by key informants related (in general) to the experiences of other coders and makers and understandings of making and coding as part of the wider field of informal activities. However, and more specifically, these accounts also reflected various positions within the making community, e.g. facilitator, activity designer, maker and programme lead. The discourse on which they drew overlapped across these different positions that were situated in both personal and professional domains.

---

[1] Learning science the fun and creative way: coding, making and play as vehicles for informal science learning in the twenty-first century https://comnplayscience.eu/.

Three interviews (each lasting approximately 30–40 min) were carried out with facilitators whom we had observed at work in the makerspace and/or those who designed the coding and/or making activities that formed part of the content observed. Three further interviews (each lasting approximately 40–60 min) were carried out with those working in similar settings to the case study site to gain further insight into the philosophies and frameworks that underpin coding and/or making activities designed for young people in informal settings. At the outset of the interview, we discussed issues around anonymity and confidentially with participants and provided each participant with an information sheet, as part of the project's consent process, approved by King's College London's Ethics Committee on 1 April 2019. Interviews were audio-recorded and transcribed soon after the interview took place. In the presented extracts, […] indicates that some text has been removed.

2.3.3   Analytical Process

Thematic Analysis (TA) is a method for analysing qualitative data that identifies patterned meaning across a dataset. Braun and Clarke's (2006) articulation of the process has been applied to a variety of disciplines and research areas. The technique has recently been further developed as Reflexive Thematic Analysis (RTA) (Braun and Clarke 2019) and is described as a subjective, organic and reflexive method of data analysis, where researcher subjectivity is understood as a resource, rather than a barrier to knowledge production. In RTA, researchers actively interpret data and create new meaning through systematic phases of research that are iterative and discursive rather than through the rigid application of a coding framework or codebook. These phases include (1) data familiarisation; (2) coding the data set; (3) generation of initial themes; (4) reviewing themes; (5) defining and naming themes; and (6) writing up the analytic narrative in the context of the literature (Braun and Clarke 2006; Clarke et al. 2015). Through this dynamic and reflective process, researchers generate new patterns of shared meaning founded upon a central concept or understanding (Braun and Clarke 2019). That is, themes do not passively emerge from the data (Ho et al. 2017).

Our data familiarisation occurred during the data collection period, through discussions during observation sessions and post-observation written reflections. Both researchers wrote individual summaries, reflections and commentaries based upon our field notes and observations within a few days of each period of observation. These notes and reflections enabled us to foreground our own subjectivities, for example, as the mothers of children a similar age to the participants, we documented our responses to what we observed as both researchers and parents; our experiences as parents undoubtedly informed our understandings of parent–child interactions in this space. On one occasion, one of us (ER) attended a making event at the case study site with her children aged 12 and 14 years. Through her participation as a parent, ER was able to observe her own children's responses to the environment of the makerspace and to the design and facilitation of activities. ER was arguably better able to draw meaning that was more contextualised from her observed responses of her own children compared to those she had never met. Relatedly, as education researchers, we brought our prior experiences of both working as a teacher in a formal classroom (ER) and as a museum educator (HK) when seeking to understand the experiences of children engaged in facilitated coding and making activities.

Our analysis process followed the phases of RTA. Meeting, on average, once per week over a 6-month period we looked for instances of where the science capital teaching approach principles were present and where they could be inserted. For example, we looked at ways in which facilitators sought to broaden what counts, or deliberately made reference to local contexts, or content that had personal significance for the participants. At the outset, we did not necessarily expect to see all eight science capital dimensions (see Fig. 2.2) enacted in the spaces we observed or described in the practices of the facilitators we interviewed. Rather, we considered particular dimensions that were more aligned with the making and makerspaces, for example, we looked for ways in which the skills inherent in the various activities were portrayed as having value and application in society (dimension 2), or whether mention was made of the various jobs and fields of study that aligned with the skills and content of the session (dimension 3). We also looked for ways in which participants were encouraged to consider making and coding activities as a normal part of their lives (dimensions 5, 8). Finally, we also noted the incidence of other practices aimed at fostering equity and social justice, and collated aspects that were relevant to our research questions. In between our meetings, we worked independently and discussed ideas via email and telephone conversation. Throughout the sequential but also recursive phases of generating initial themes, reviewing themes, and defining and naming themes, we sought to generate new understandings of the ways in which informal making and coding activities and spaces were, or could be, socially just and equitable. Therefore, our analysis was deductive (i.e. directed by existing ideas, in this case the concept of science capital) and latent (i.e. reporting concepts and assumptions underpinning the data) and situated in our familiarity with the SCTA (Godec et al. 2017).

2.4 Findings

Our research was guided by the question:

To what extent do makerspaces engender equitable access to the STEM learning opportunities inherent in the space?

In keeping with our theoretical lens of the science capital teaching approach, we examined the data with respect to the extent that the practice broadens what counts as STEM learning. We were conscious that broadening what counts in the context of a makerspace may look quite different from practices enacted in school settings. We looked for ways in which the young learners were made physically comfortable in what for many was a large, unfamiliar, albeit exciting, space. One facilitator, Ian, gave children gentle, positive reinforcement, speaking to them in a soft and encouraging tone and, in contrast to other facilitators who seemed more comfortable leading from the front of the space, frequently sat side-by-side with the children as they built robots or completed coding activities. As a result, children who had initially appeared unable or unwilling to contribute happily volunteered to share their ideas with him in person and through group discussions. For example, during a question and answer session with 12 children (two of whom are girls) led by another facilitator Marc, we observed the following in our field notes:

> At one point during a discussion session in the sofa area, Ian is sitting with the children whilst Marc leads the Q&A. Ian is seated near a more reluctant female participant and he encourages her to raise her hand and share her idea with the group. It is hard to hear her voice, and so Ian repeats her idea warmly and with enthusiasm. After Ian encourages the first, the second girl in the group volunteers an idea to the group.

In addition, Ian worked hard to frame his contributions in ways that younger learners would understand. He soon discovered that references to the film *Jaws* were not successful and instead sought to find a more contemporary film reference to highlight the use of robots and animatronics to create responses with audiences:

> Ian has a group discussion with the children about robots and how they can create emotions in audiences, for example in films. He hums the theme music to *Jaws* as a prompt, but this only receives blank faces! He quickly responds and asks the children which robots they have seen in films or on TV and this sparks a discussion about robots and transformers.

In this way, Ian remained open to children's experiences and knowledge and did not exhibit rigid expectations of what constitutes norms in knowledge and behaviour. Thus, in his approach to facilitation, Ian's physical behaviour and his affirmative ways of speaking were seen to be an example of broadening what counts: the atmosphere he engendered created opportunities for the learners to express themselves in ways they felt comfortable and, through his use of the young people's contributions, Ian promoted a culture of respect.

Other facilitators had a different approach to facilitation which could be characterised as energetic and enthusiastic with a strong physical presence. For example, Marc chose to wait at the entrance to the makerspace to welcome children and their parents, using a loud and enthusiastic voice, creating an atmosphere of excitement and energy. In the makerspace, Marc's questioning style was ebullient, energetic and dynamic, with questions coming in rapid succession:

> Marc begins by asking the children: what kind of sensor would we need to detect sound? What is taste? Where is taste detected? How many flavours are there? These questions are asked using a loud voice with Marc standing at the front whilst the children are sat on the floor at his feet. Marc asks children to put their hands up to respond and some of them do.

These questions are all well directed in terms of content, and Marc's enthusiasm for the topic is clear: we observed some children positively engage with Marc's style, expressing their excitement and engagement with loud affirmative responses when asked 'are you ready to make?' However, we also noted that Marc's pace was extremely fast, and that some children did not volunteer to answer any of the questions, perhaps because they were not able to respond in the time available, or did not feel confident enough in the setting to put their hand up.

However, it is important to note that Marc also employed other, quieter, approaches during his interactions with young people. At a later point in the session, he used a visualisation technique to encourage participants to use their imaginations in the development of their making designs:

> […] imagine your creations today…it could have wheels…it could have wings…Which sensors? Maybe a name? Try to visualise it in your brain.

During this activity, the children shut their eyes and listened carefully. Then, when Marc invited them to begin the maker challenge at the end of the visualisation activity, all did so in a highly focused manner.

There were frequent opportunities for facilitators to personalise and localise the content by drawing on the experiences of participating young people. During one session, children were asked to think of robots that would be helpful to them in their homes and lives. The children were encouraged to think of ways in which technological solutions could be of value to their parents and/or carers. In our field notes, we noted the following:

> At the beginning of a session a facilitator showed a short, amusing video-clip of a robot doing household tasks e.g. hoovering. The facilitator then asked the children, 'what would make your life at home easier?' 'What could you create to help your parents?' The children offered up a variety of examples including a drinks shaker, a clothes dispenser, a robot that plays the piano and a robot with a fan. The facilitator described these as good examples of 'life hacks' that would help their parents and make their lives easier.

Here, the facilitator supported the children to draw on experiences from their own lives and contexts, and explicitly acknowledged these as having relevance both to the children and also other significant adults in their lives. In another activity where children were using a motor to create a moving 'insect', facilitator Jen described how she used 'everyday examples' including a fan, a microwave and an electric toothbrush to encourage children to recognise that motors were something that were part of their lives and relevant to them. In this way, Jen was personalising and localising the core concept of the motor for the children.

We observed distinct approaches to the use of materials such as using craft materials, cardboard and Lego with which the children were familiar and felt comfortable. We also noted many comments in the sessions encouraging children to build similar makes with materials found in their homes. One facilitator, Pete, led a discussion about cardboard when he shared an example of a robot he had largely made using recycled cardboard:

> As Pete shows the children the robot, he tells the children that cardboard is an 'everyday material' and asks the children to share some examples of how they have used cardboard in the last few days. The children respond with examples including, cereal boxes, toilet roll tubes and toothpaste boxes. Ian replies, 'Yes, that is great, you see I have used a Rice Krispies box for the body of my robot, you don't need to use fancy materials, and I bet you have lots of ideas of how to create a robot from what other people might see as junk.

In this way, Pete demonstrates that ordinary materials can be reused to create exciting new things. Furthermore, in using an everyday object such as a cereal box (and a brand that is primarily aimed at children) in his model of a robot, he is using examples that are very familiar and accessible to children. The impact of localising materials in this way is enhanced by Pete holding the robot in his hands: the children are presented with a tangible example of how materials with which they are familiar can be re-imagined and reformed.

We looked for ways in which young peoples' ideas were elicited and valued, including instances where their prior experiences of making and coding were noted and built upon. Our observations suggest that facilitators regularly encouraged children to share their ideas drawn from their own experiences and affirmed the value of their ideas as a legitimate source of knowing. This was predominantly achieved through group question and answer sessions that were initiated and led by the facilitators. Some facilitators were able to ensure children's contributions were recognised by other adults and children, by facilitating conversations between children and between children and facilitators. In one case, a facilitator (John) made verbal connections between the contributions of children who felt able to share their experiences in response to a question and those who the facilitator had observed during the previous activity:

> A boy explains how he made a robotic insect during a group question and answer session. In response to his contribution, John took the opportunity to make a link between the boy's creation and that of two girls who John had supported during the previous activity. John highlighted the children's knowledge of biology in designing the accurate insect-shaped robots and explicitly valued their creations and their contributions through the group discussion.

In this way, John is eliciting a more diverse range of contributions by inviting children to respond to questions and then taking those contributions and linking them to the work of other children he had observed. This also creates a collaborative and supportive atmosphere, where the efforts of all are recognised and valued. Over the course of a day-long activity children gravitated towards the facilitators who regularly, and meaningfully, valued their efforts and forged positive connections between children. Children who felt able to share their thoughts and ideas with their peers and the facilitators were praised and had their prior experience affirmed.

We noted instances when key science capital dimensions were used and when they were not included. We observed frequent opportunities for the development of children's scientific literacy (dimension 1). For example, the use of 'wildlife' as the overarching theme provided the framework for a robot-building activity. The facilitator, Edward, encouraged children to consider the wildlife they had seen and through this think about the concept of wildlife:

> Edward asks the children, 'what wildlife have you seen in the city?' Children respond, 'dogs, cats'. Edward says, 'are they wild?' One child responds that these are pets, but that lemurs and tigers are wild. Edward asks the children if they have seen tigers moving around the city? The children laugh and say they haven't and when Edward asks them again to give examples of wild animals they have seen in the city around them they give examples which include pigeons, ducks and foxes.

Edward then extends this discussion of wildlife by asking the children about the challenges that these animals face. Edward said: 'we want our city to be inclusive, we need to be mindful of that, so what can we do?' The children share ideas about clean rivers and parks that are litter-free and Edward encourages them to use terms including 'pollution' and 'recycling'. In this way, Edward is building the children's scientific literacy by using accurate scientific terms and through questioning, drawing out children's misconceptions and supporting them to developing their understanding of what wildlife means in the context of where they live.

The selection and use of resources also afford the development of young people's science media consumption (dimension 4). For example, when explaining a challenge, facilitators used short multimedia clips of a robot alarm clock that were amusing and engaging:

> After a break, Rosie gathers the children to the mat and shares with them an animation of an alarm clock that appears normal, but when it goes off, arms spring out from each side and as the alarm clock rings it also makes a cup of coffee for the sleepy person which causes the person to fall out of bed in surprise. The children giggle and laugh. The facilitator then says, 'well that robot didn't go quite to plan did it? What robot could you create that would be useful?

The animation and other short clips of real robots being used in the home served as a way of focusing participants' attention on the task in hand: developing practical technology-based solutions to everyday social problems. The children enjoyed these clips and requested that they be played during other periods of social time (e.g. lunchtime) and facilitators shared information about how to access them at home. In this way, children were able to use the resources to extend their science and technology consumption at home and potentially share this with their family and friends, thus providing a further opportunity to widen their experience of science with key people in their lives.

The recruitment of facilitation teams resulted in groups of facilitators that were often diverse and reflected the socio-economic and cultural diversity of the participating children. Children called facilitators by their first names, and, during small group work, were able to have lengthy conversations with facilitators. As a result, some children forged a rapport with these adults, and learnt about their motivations and interests as the following field notes highlight:

> Jen is demonstrating how to make a moving bug-like creature using a motor, a CD and craft materials. She says that she found it tricky to work out how to make the microbit work in the way she wanted to but that she kept trying and asked for help, and that this is important when you are a maker. She says you need to 'try, try, try again when things get hard'. She explains how she enjoyed choosing the different craft materials for her bug and how it was like a character from a story book that she used to read as a child.

Here, Jen explicitly shared how she had drawn on her own childhood experiences and interests when developing her bug. Jen also shared the implicit message that, through participation in such activities, children become makers and that their own experiences and interests are valuable. Jen also foregrounded that makers regularly encounter difficulties, but that this is to be expected. Furthermore, if they persist, makers can be successful. By sharing her identity as a maker, Jen essentially extended the children's network of people who are involved in STEM-related jobs and hobbies. Significantly, she also signalled that being a maker was something the children could remain being once they were older.

2.5   Discussion

Having explored and documented the ways in which the practice of makerspaces and makerspace facilitators can readily align with the principles of science capital, we turn now to consider the ways in which such practices could be amended and further refined.

Firstly, we underscore the importance of reflecting on the varied styles employed by facilitators and note the need for greater flexibility in approach to ensure that all participants feel able to contribute. For example, Ian's softly spoken and gentle approach markedly contrasted with Marc's more ebullient style. The former approach had the effect of enabling children to see their contributions as valuable, and to learn that encountering, and then addressing, difficulties were part and parcel of the maker/coder experience. The latter approach proved to be particularly engaging to those children who already possessed a degree of self-confidence, and/or previous experience of coding or making. Whilst both approaches have their value, care must be taken to ensure that children are supported—especially those who may not be confident with the rules of the makerspace. In short, an approach to facilitation that actively seeks to broaden what counts will serve to challenge any tendency for makerspaces to perpetuate unreflective practices. It will also prevent the creation of a space in which the facilitator has all the power and voice, and where learners' needs and preferences are secondary to the expectations of adults.

Secondly, whilst we noted facilitators regularly eliciting information from the children about their previous experience of coding and making at the outset of an activity, we rarely observed facilitators building upon these prior experiences and referencing the children's examples thereafter or incorporating them into subsequent activities. We believe that facilitators can do more to value children's contributions and explicitly recognise the resources they bring to the session from their prior life experience.

Thirdly, although some facilitators used everyday materials and examples of resources to personalise making, these examples of personalising were the exception rather than the rule. For example, although facilitators were at times able to use questioning to reveal and correct children's misconceptions, a greater emphasis on the lived experiences of children could have enabled more equitable engagement. This would provide a focus that relies on what children have experienced rather than what they may or may not already know.

Fourthly, through our observations we saw instances of science capital dimensions. In particular, we noted the inclusion of science literacy (dimension 1) and science media consumption (dimension 4). However, we argue that much more could be done to include more dimensions more often in ways that become regular and intentional parts of practice. For example, during the workshops we observed, facilitators did not always introduce themselves to the children, and when they did, it was simply to identify themselves by their names. Furthermore, rarely did they introduce themselves as having any valid knowledge of making or coding and yet most facilitators (some of whom were volunteers from local tech-based businesses) had extensive experience of careers in STEM, including professional roles in programming, making and design. Over the course of a day-long, out-of-school experience, children built a rapport with some facilitators and clearly saw them as role models when completing activities (dimension 7). Had facilitators shared their wider identities as programmers and makers and highlighted the range of occupations that can ensue from an understanding of, and interest in, making and coding (dimension 3), children may have viewed the facilitators not only as short-term mentors but as role models for their future lives.

2.6 Concluding Thoughts and Future Directions

We have demonstrated that science capital pedagogical principles are utilised in makerspaces (even if they look a little different from schools), but that more is needed to ensure that these learning environments fulfil their potential. We have argued that science capital approaches are effective in creating an environment where young people feel valued, comfortable and therefore more likely to engage in making and coding. Nonetheless, we note a wider body of research from the US (Lewis 2015; Kim et al. 2018) and the UK (Dawson 2017) that suggests making and makerspaces are faced with a continuing equity issue, with boys being more likely to prosper than girls. In discussing our findings, we have pointed to the effect that even small changes in practice can have in ensuring that makerspaces can be more equitable and socially just experiences. However, we also recognise that the lens of science capital may not fully address the complexity of practices inherent in the informal, facilitated learning environments of makerspaces: further theorisations may be needed. Indeed, in our ongoing work, we are exploring the role of parents as those who support children's participation and engagement in makerspaces. We are also reflecting on the framings of makerspaces as playful and are considering how appropriate this is as a positioning of the learning that is found therein.

Acknowledgements The authors would like to thank the staff and children of the case study site for their support and contribution to this research. This project has received funding from the European Union's Horizon 2020 research and innovation programme under grant agreement NO 787476. This paper reflects only the authors' views. The Research Executive Agency (REA) and the European Commission are not responsible for any use that may be made of the information it contains.

References

Archer, L., DeWitt, J., Osborne, J., Dillon, J., Willis, B., & Wong, B. (2013). 'Not girly, not sexy, not glamorous': Primary school girls' and parents' constructions of science aspirations. *Pedagogy, Culture & Society, 21*(1), 171–194.

Archer, L., Dawson, E., DeWitt, J., Seakins, A., & Wong, B. (2015). "Science capital": A conceptual, methodological, and empirical argument for extending bourdieusian notions of capital beyond the arts. *Journal of Research in Science Teaching, 52*(7), 922–948.

Archer, L., DeWitt, J., & Osborne, J. (2015). Is science for us? Black students' and parents' views of science and science careers. *Science Education, 99*(2), 199–237.

Blikstein, P. (2013). Digital fabrication and 'making' in education: The democratization of invention. In J. Walter-Herrmann & C. Büching (Eds.), *FabLabs: Of machines, makers and inventors*. Bielefeld: Transcript Publishers.

Bourdieu, P. (1977). *Outline of a theory of practice* (Vol. 16). Cambridge, UK: Cambridge University Press.

Bourdieu, P. (1986). The forms of capital. In J. G. Richarson (Ed.), *Handbook of theory and research for the sociology of education* (pp. 241–258). New York: Greewood Press.

Brahms, L., & Crowley, K. (2016). Making sense of making: Defining learning practices in MAKE magazine. *Makeology: Makers as Learners, 2*, 13–28.

Braun, V., & Clarke, V. (2006). Using thematic analysis in psychology. *Qualitative Research in Psychology, 3*(2), 77–101.

Braun, V., & Clarke, V. (2019). Reflecting on reflexive thematic analysis. *Qualitative Research in Sport, Exercise and Health, 11*(4), 589–597.

Braun, V., Terry, G., Gavey, N., & Fenaughty, J. (2009). 'Risk' and sexual coercion among gay and bisexual men in Aotearoa/New Zealand–key informant accounts. *Culture, Health & Sexuality, 11*(2), 111–124.

Calabrese Barton, A., Tan, E., & Greenberg, D. (2017). The makerspace movement: Sites of possibilities for equitable opportunities to engage underrepresented youth in STEM. *Teachers College Record, 119*(6), 11–44.

Clarke, V., Braun, V., & Hayfield, N. (2015). Thematic analysis. In J. A. Smith (Ed.), *Qualitative psychology: A practical guide to research methods* (pp. 222–248). London: Sage.

Dawson, E. (2017). Social justice and out-of-school science learning: Exploring equity in science television, science clubs and maker spaces. *Science Education, 101*(4), 539.

DeWitt, J., Archer, L., Osborne, J., Dillon, J., Willis, B., & Wong, B. (2010). High aspirations but low progression: The science aspirations–careers paradox amongst minority ethnic students. *International Journal of Science and Mathematics Education, 9*(2), 243–271.

DeWitt, J., & Archer, L. (2017). Participation in informal science learning experiences: The rich get richer? *International Journal of Science Education, Part B, 7*(4), 356–373.

DeWitt, J., Nomikou, E., & Godec, S. (2018). Recognising and valuing student engagement in science museums. *Museum Management and Curatorship, 34*(2), 183–200.

Godec, S., King, H., & Archer, L. (2017). *The Science Capital Teaching Approach: Engaging students with science, promoting social justice*. London: University College London.

Godec, S., King, H., Archer, L., Dawson, E., & Seakins, A. (2018). Examining Student Engagement with Science Through a Bourdieusian Notion of Field. *Science & Education, 27*(5–6), 501–521.

Halverson, E. R., & Sheridan, K. (2014). The maker movement in education. *Harvard Educational Review, 84*(4), 495–504.

Ho, K. H., Chiang, V. C., & Leung, D. (2017). Hermeneutic phenomenological analysis: The 'possibility' beyond 'actuality' in thematic analysis. *Journal of Advanced Nursing, 73*(7), 1757–1766.

Honey, M., & Kanter, D. (Eds.). (2013). *Design, Make, Play: Growing the next generation of STEM innovators*. London: Routledge.

Kim, Y.E., Edouard, K., Alderfer, K. & Smith, B.K. (2018). *Making culture. A National Study of Education Makerspaces*. ExCITe Centre Report. Retrieved from: https://drexel.edu/excite/eng agement/learning-innovation/making-culture-report/

Lewis, J. (2015). *Barriers to women's involvement in hackspaces and makerspaces*. Access as spaces. Available at: https://access-space.org/wp-content/uploads/2015/10/Barriers-to-womens-involvement-in-hackspaces-and-makerspaces.pdf. Accessed May 10, 2016.

Martin, L. (2015). The promise of the Maker Movement for education. *Journal of Pre-college Engineering Education Research* (J-PEER), *5*(1), 30–39.

Martin, L., Dixon, C., & Betser, S. (2018). Iterative design toward equity: Youth repertoires of practice in a high school maker space. *Equity & Excellence in Education, 51*(1), 36–47.

Sheridan, K., Halverson, E. R., Litts, B., Brahms, L., Jacobs-Priebe, L., & Owens, T. (2014). Learning in the making: A comparative case study of three makerspaces. *Harvard Educational Review, 84*(4), 505–531.

Heather King is Reader in Science Education at King's College London. She has long been involved in researching teaching and learning in out-of-school settings such as museums and galleries. Her most recent work has involved longitudinal studies of secondary science classrooms with the aim of better understanding equitable practice and supporting more young people feel able to engage in STEM.

Elizabeth A. C. Rushton has worked within education as a high school teacher, as Director of Evaluation for an education charity that supports school student participation in STEM research and is currently a Research Associate at King's College London. Her research considers young people's experience of science in formal and informal settings and teacher professional development through collaborations with researchers and mentoring school student research.

Part II
Technological Frameworks, Development and Implementation

This part provides insight into different technologies and their ability to enhance science learning.

Chapter 3
Digital Games for Science Learning and Scientific Literacy

Iro Voulgari

Abstract In this chapter, we focus on the links between science learning and digital games. We review previous studies in the field, identify key findings and propose a conceptual model for further research. We view digital games not only as media through which players can explore and understand or be motivated to further study the learning content, but also as cultural and social practices within which gameplaying is situated. The main themes discussed in this chapter as factors relevant to the support of science learning and scientific thinking through game-based learning are (a) game design issues, (b) individual factors such as game preferences and motivations, game experience and literacy, and perceptions of games and (c) the social and cultural context of gameplaying (e.g. formal, non-formal and informal learning settings). Digital games can be effective instructional tools for science education but in this chapter we further examine how they can become tools for empowering the learners to meaningfully engage with science and how they can support the learners' scientific literacy and citizenship.

Keywords Science learning · Scientific literacy · Game-based learning · Digital games · Literature review

3.1 Introduction

Back in 1997, in his book *"The demon-haunted world: science as a candle in the dark"*, Carl Sagan wrote about the importance of scientific thinking and the scientific method in our everyday lives, and how crucial critical and sceptical thinking against fallacious arguments and deception is. Over the past years, with the spread of disinformation and the role the media and their impact on people's behaviours, decisions and attitudes (Koltay 2011), the importance of critical and scientific thinking is still relevant. Skills for evaluating evidence-based claims such as news articles and advertisements and for identifying *"bogus claims"* are needed for everyone and

I. Voulgari (✉)
Institute of Digital Games, University of Malta, Msida, Malta
e-mail: iro.voulgari@um.edu.mt

© Springer Nature Singapore Pte Ltd. 2020
M. Giannakos (ed.), *Non-Formal and Informal Science Learning in the ICT Era*, Lecture Notes in Educational Technology,
https://doi.org/10.1007/978-981-15-6747-6_3

particularly for children as future consumers, scientists and citizens (Halpern et al. 2012).

In this chapter, we focus on the links between science learning and digital games. We review previous studies in the field, identify key findings and propose a conceptual model for further research. We view digital games not only as media through which players can explore and understand or be motivated to further study the learning content, but also as cultural and social practices within which gameplaying is situated.

The main themes discussed in this chapter as factors relevant to the support of science learning and scientific thinking through game-based learning are (a) game design issues, (b) individual factors such as game preferences and motivations, game experience and literacy, and perceptions of games and (c) the context of gameplaying (e.g. formal learning settings, non-formal settings such as workshops on game-based learning, informal settings such as game exhibitions and contests).

3.1.1 Scientific Literacy

Scientific literacy involves a range of skills and concepts such as science identity, scientific reasoning, scientific enquiry and mastery with science-related content, activities and methods. These skills and concepts relate not only to content knowledge, but also to knowledge and understanding of scientific practices, as well as the global context science is situated in and its contribution to society (Fraser et al. 2014; OECD 2012; Wallon et al. 2018). Knowledge of the scientific practices and content knowledge seem to complement each other. Science content knowledge is related to scientific sense making and scientific literacy skills (Cannady et al. 2019). Students who engage with scientific practices can learn science content more effectively.

A number of different factors seem to have an impact on the scientific literacy of children. Archer et al. (2015) discussed the concept of *science capital* as a set of environmental factors affecting students' attitudes towards science, such as their family, friends and daily activities (e.g. visits to museums, after-school programmes, access to science-related resources). Similarly, Markus and Nurius (1986) discussed the concept of the *possible self* (i.e. expectations and hopes of what one can become in the future) and linked it with personal experience and the environment, e.g. cultural norms, friends, teachers, parents and the media. Beier et al. (2012) built upon the *possible self* construct and proposed a measure for the scientific *possible self* of students for examining the impact of a science-focused game on the possible selves of the students.

In a formal or informal learning environment, interesting and motivating experiences may have a positive effect on personal interest and engagement with science-related activities, as well as on individual attitudes and predispositions towards science. Studies have shown that instructional interventions and techniques triggering *situational interest*, such as hands-on activities, toys and science games in formal education settings can increase *individual interest* and further participation in science-related activities in informal settings, such as talking, thinking and reading about science (Hidi 1990).

It seems, therefore, that science literacy and attitudes towards science are influenced by the quality of the learning activities as well as the context these activities are situated in, and the social environment of the children.

3.2   Game-Based Learning

Extensive literature on game-based learning over the past 20 years shows that games can be used as instructional and learning tools; they can integrate various learning theories and pedagogical techniques; they can support a number of different learning outcomes such as content understanding and problem-solving skills and facilitate transfer of knowledge and understanding of processes and practices to other domains (Egenfeldt-Nielsen 2007; Kafai 2006; Ke 2009; O'Neil et al. 2005; Shaffer 2006).

As digital games can simulate complex systems and allow the players to explore and experiment with the role of each component and the relationships among the components, they have the potential to support scientific literacy objectives (e.g. content knowledge, systems thinking, social implications). In their report, Clark et al. (2009) review existing games and studies on science learning, and identify goals such as conceptual understanding and process skills, epistemological understanding, attitudes and identity, and design issues. Games can further be motivating learning experiences for the students, increasing the depth and duration of the students' engagement with the learning content (Cordova and Lepper 1996; Ryan et al. 2006). Considering this potential of games for science learning (National Research Council 2011, p. 2), and the need for further study of the factors involved (see also Li and Tsai 2013 for a meta-review on this topic), we reviewed latest literature in order to identify trends and factors in relation to the games' content, design and integration into learning settings for science literacy.

3.3   Science Learning and Digital Games

Although this chapter is not an extensive, empirical meta-review, we tried to get a better and unbiased understanding of the area by following a more standardised protocol for identifying representative trends in the area: we used Google Scholar as a publicly available index of scholarly literature with the keywords ("games" OR "game") AND ("science" OR "scientific"), so that the search can be easily reproduced. The search, on October 2019, returned approximately 994 results without including patents or citations. After (a) limiting the range to more recent studies, over the past decade following up on Li and Tsai's review (2013) and up to 2019, (b) excluding papers not written in English, (c) only including journal articles which had received at least 10 citations for ensuring the quality of the studies reviewed and

(d) excluding papers not relevant to digital games and science education or learning, the remaining set of 30 papers was more thoroughly examined. In the following sections, we describe main trends relevant to the learning goals, the games used, factors involved, research settings and methods.

3.3.1 Research Methods

Most of the studies reviewed used experimental or quasi-experimental conditions, collecting data from pre-post surveys measuring constructs such as flow experience, knowledge, attitudes about games and perceptions about self-efficacy in science or games. Observations and interviews have also been used for collecting data on the context of gameplay and for gaining more in-depth insights on the motivations, perceptions and interpretations of the participants. The participants' concept maps have also been used as data collection instruments for analysing their perceptions, mental schemata and prior knowledge [e.g. Waddington and Fennewald (2018)].

A large number of studies further examined the actual gameplay, in-game behaviours and performance of the participants, collecting and analysing data such as video recordings of gameplay, game metrics such as playtime and number of restarts, log files, eye-tracking and videos of facial expressions for studying attention allocation and emotion (Ault et al. 2015; Hou 2015; Muehrer et al. 2012; Taub et al. 2018).

In one case, where physical activity was also examined, heart rate monitors were used for data collection (Sun and Gao 2016). An ethnographic study was used, in another case, on public online fora, for studying motivations for participation in citizen science projects (Ponti et al. 2018).

3.3.2 Science Domains and Learning Objectives

Recent reports and meta-reviews of empirical studies on game-based learning for science learning indicate an emphasis on learning goals such as learning scientific content knowledge, conceptual understanding and knowledge, knowledge construction, problem-solving, engagement and participation, while aspects such as complex problem-solving, critical thinking, understanding of scientific processes, epistemological understanding, the potential of games to motivate interest in science, affective outcomes, and socio-contextual learning are less researched (Cheng et al. 2015a, b; Li and Tsai 2013; Martinez-Garza et al. 2013; National Research Council Report 2011, p. 2). It was also found that the games used in previous studies were mainly focused on physics and biology or they were interdisciplinary. With these limitations in mind, we sought to examine whether these trends persisted over the past decade and also identify any relevant factors as barriers or possibilities.

Most of the studies we reviewed focus on scientific fields such as Physics, Biology, Chemistry and Environmental Education, with very few examining game-based learning in Social Sciences. Additionally, there seems to be a shift to learning objectives such as understanding of scientific processes and practices, attitudes towards science and higher order thinking skills.

Biology-related games (e.g. on neuroscience, virology, evolution), mainly single-player except Ketelhut et al. (2010) who used the multiplayer game *River City*, featured quite prominently among the studies on science learning, examining learning outcomes such as scientific argumentation, scientific inquiry (e.g. making hypotheses, gathering and analysing data, proposing predictions), conceptual understanding of scientific processes, transfer of knowledge, procedural knowledge and higher level of cognitive process, with generally positive results (Bergey et al. 2015; Cheng et al. 2014; Wallon et al. 2018). Israel et al. (2016) examined the relation between personal characteristics such as learning disability, gender and perceptions of games with the learning outcomes, situated their study in the context of scientific literacy, informed citizenship and interest in science-related careers and developed biology games (*Cell Command, Crazy Plant Shop, You Make Me Sick!*) aiming to address both content knowledge and problem-solving ("*thinking like a scientist*"). Although Marino et al. (2013) didn't focus on learning outcomes but rather on the correlations among factors such as gameplaying behaviours, reading ability, perceptions on scientific ability and disability status, they used the games *You Make Me Sick!* and *Prisoner of Echo* which focus on virology and physics, respectively, for examining students' attitudes about science and the work of scientists, and learning science through games. Even though Cheng et al. (2015a, b) mainly addressed content knowledge using the game *Virtual Age*, they recognised the need for further study of learning outcomes such as problem-solving and scientific reasoning. Results were not always positive, though, depending on certain conditions. Muehrer's et al. (2012) results, for instance, who used the game *Genomics Digital Lab*, showed that students improved their science vocabulary and not their understanding of abstract concepts. Also, Taub et al. (2018), using the game *Crystal Island*, found that efficiency at solving the game problems was significantly related to the gaming behaviours of the players (e.g. manipulation of the game items), concluding that appropriate scaffolding is required in game-based learning environments.

Physics (e.g. light and shadow, Newtonian mechanics, the solar system) was another prominent science domain for science learning through games. Similarly, higher order cognitive skills were addressed, such as scientific knowledge construction (Hsu et al. 2011), conceptual change—using the game *Space Challenge* (Koops and Hoevenaar 2013), scientific inquiry through experimentation and collaborative learning—with *Quantum Moves* (where students had to build a quantum computer) (Magnussen et al. 2014), content knowledge, problem-solving and scientific inquiry—with *Alien Rescue* (Liu et al. 2014), and also implicit science knowledge—with the *Carrot Land* (Chen et al. 2015) who also observed the emergence of collaboration and collaborative problem-solving in the collaborative play condition. Sun and Gao (2016) combined the physics game *Earth, Moon and Sun*, where students have to learn information about the solar system, with a stepper for students to control the game with, for examining science learning and motivation, in relation to physical activity. In both conditions, they found increased learning outcomes (science knowledge) and situational interest.

Similar learning objectives, such as science content, problem-solving, scientific inquiry meta-cognitive processes, scientific argumentation, motivation to engage in science and systems thinking, were studied in fields such as map reading (corresponding to STEM education-related objectives in the United States curriculum) using the game *Crystal Island* (Lester et al. 2014), chemistry with the game *Perfect PAPA II* in relation to the learners' flow experience and behavioural patterns (Hou 2015), STEM-related games (Schifter et al. 2012), the multidisciplinary game *Reason Raser* relevant to "earth and space, life, physical, and technology and engineering sciences" (Ault et al. 2015). Again the outcomes on students' performance, confidence and motivation to engage in science were positive, but under certain conditions. Waddington and Fennewald (2018), for example, used a climate change simulation game (*Fate of the World*). Their results were promising for the development of deeper and more robust systems thinking but learning outcomes and game interpretations by the players were limited due to the design and the mechanics of the game.

There were fewer studies focusing on Social Sciences. The study of Sáez-López et al. (2015), for example, examined games for teaching Social Sciences in the classroom and identified a number of games of an "*economic, social, geographical, artistic and historic nature*", while Lee and Probert (2010) examined the game *Civilization III* for Social Studies teaching (History) to high school students, and highlighted the decision-making and problem-solving processes the students engaged in during gameplay, as well as the content knowledge they acquired. In their review (VanFossen et al. 2009) discussed the potential of Massively Multiplayer Online Role-Playing Games (MMORPGs) as learning tools for citizenship education in the social studies classroom; students potentially experience teamwork, understanding and tolerance of others, practice decision-making skills, and they can be encouraged to reflect on and discuss issues such as governance, rights and economic principles.

Games seem to have the potential to link the game experiences with a wider social and global context by strengthening the students' scientific identities and awareness of real-world problems. For instance, Marino and Hayes (2012) argued in favour of the potential of appropriately designed games to enhance science education, civic scientific literacy and participation of students in scientific discourse, referencing relevant empirical studies and games such as *River City*, *Quest Atlantis* and *Whyville*. Gaydos and Squire (2012) studied the game *Citizen Science* in school settings in relation to the students' identities as citizen scientists. The goal of the games was to "*encourage democratic participation in society by providing students with the perspective that they are capable of acting as legitimate sources of science-driven community activism.*" The scientific identity of 13–14-year-old students was also strengthened in Chee and Tan's (2012) study. The game used (*Legends of Alkhimia*) was an educational game about chemistry and through its inquiry-based design it helped students not only develop their understanding of chemistry but also engage in scientific processes such as critical thinking and experimentation, and positively enhance their perceptions for their scientific identities and their dispositions towards

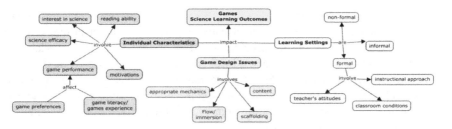

Fig. 3.1 Conceptual model of factors found to be relevant to science learning outcomes through digital games

science. Beier et al. (2012) examined the impact of a science-focused game on the scientific *possible selves* of middle-school students; their findings suggested that the game had a positive impact on students' acquisition of science content, science process skills and also their motivations for careers in science. In a wider social context, Chee and Tan (2012) and Dippel and Fizek (2019) discussed the role of playful technologies and games, situated in a context of collaboration and an external purpose, as tools for engaging in citizen science. Indeed, scientific ideals and also fun were the main motivations of people for participating in citizen science games (*Foldit, Galazy Zoo*) as identified in Ponti et al. (2018) ethnographic study of the public forums of the two games and also by Curtis (2015) who identified contribution, interest in science, interactions with others and challenge as the main motivations for participation in the online citizen science game *Foldit*.

Very few of the studies and games adopted a more multidisciplinary approach integrating multiple domains, even though games are appropriate environments for such approaches. Only in one case, the study addressed critical thinking, scientific reasoning and transfer of knowledge, across different domains of knowledge (psychology, biology and chemistry) using the game *Operation ARA (Acquiring Research Acumen)* (Halpern et al. 2012). This could be attributed either to research design goals (e.g. study-specific learning outcomes) or to the fact that a game used in school settings will have to comply with specific curriculum goals and final tests requirements such as explicit learning outcomes, for justifying its role and for being easier accepted by the teachers (Magnussen et al. 2014).

As previously implied, the efficiency of science literacy learning through games is related to certain factors. These factors are relevant to learner characteristics, game design and settings of the gameplay sessions and will be discussed in the following sections (see also conceptual model in Fig. 3.1).

3.3.3   Individual Characteristics

In their study, Fraser et al. (2014) examined the associations amongst youths' science identity, science understanding and gaming preferences, and identified personal game preferences as an important factor for the effectiveness of games in science learning. Different types of games and different game features may attract different types of learners. Social gamers, for instance, may be attracted by in-game social and instrumental interactions with others and seek support from others. Lower motivation, immersion and flow experience in the game may lead to fewer learning behaviours (Hou 2015). Science literacy games failing to consider the preferences of the learners may, therefore, produce lower learning outcomes, at least for the learners who are not attracted by the type of the game.

Previous game experience was one of the mediating factors for the learning outcomes, in the studies reviewed. Game literacy and understanding of the game conventions lead to better content understanding and learning performance as observed by Gaydos and Squire (2012). Computer game self-efficacy, on the other hand, does not seem to significantly predict performance (Bergey et al. 2015). In Waddington and Fennewald (2018), previous game experience was a prerequisite for participation and even so, players found it difficult to navigate the affordances of the simulation game. On the other hand, based on findings in Bergey et al. (2015), game self-efficacy may not significantly predict performance in the game, and performance did not significantly predict changes in game self-efficacy. Succeeding in the game may not necessarily mean that the students achieved a better conceptual understanding (Muehrer et al. 2012). The students may focus on the game mechanics and discuss how to win the game, without necessarily gaining a deeper understanding of the content. And, in addition, players may develop their own meanings and interpretations of the game, sometimes entirely different from those intended or anticipated by the game developers (Waddington and Fennewald 2018). Game literacy, therefore, is an important factor for the learning effectiveness of the game but does not guarantee it.

Previous interest in science and the academic performance in science-related subjects of the learners may also impact the gaming performance and the learning outcomes of a science game. Factors such as reading ability, prior knowledge on the topic, perceptions about science knowledgeability and initial scientific inquiry self-efficacy influence the gaming performance and achievement, and the changes in scientific inquiry self-efficacy (Bergey et al. 2015; Israel et al. 2016). In addition, games involving science learning are more likely to be preferred by students with already high academic performance and science literacy, and students with a higher science literacy level may already spend more time in playing science games out-of-school (Fraser et al. 2014). In Taub et al. (2018), students who recognised and manipulated the in-game items that were more relevant to the problem managed to solve the problem more efficiently, which may be related to the familiarity of the learners with the game content. The science content of the game may also affect learning effects of video games on science learning (Israel et al. 2016). It seems, thus, that science-related games, at least those tested in the studies reviewed, may attract and benefit students with an already high level and interest in science-related topics, while excluding the students who actually need them more.

Gender does not significantly predict game performance and learning outcomes. In Bergey et al. (2015), although girls had lower scores in game self-efficacy, this did not seem to affect their game performance. Girls had also lower scores in attitudes about learning from games, but they had no other significant differences with boys on science achievement perceptions and interest in careers in science (Israel et al. 2016).

3.3.4 Design Issues

Drawing from studies indicating the importance of personal preferences in games and the importance of flow and immersion for the learning outcomes, it seems critical that the game design addresses these aspects; games adapted or adapting to the learner or player type, scaffolding participants based on in-game behaviours, increasing immersion and flow state of the players through elements such as clear goals and immediate feedback, have been proposed as design guidelines for effective science learning games (Cheng et al. 2015a, b; Hou 2015; Taub et al. 2018).

Certainly, designing immersive, engaging and adapted to the target group's requirements is not enough for science learning; the design of the learning content is equally critical. Clark et al. (2015) discussed the importance of designing games where learners can interact with models and systems accurately conveying the science content, phenomenon, system, model and relationships involved, allowing them to further engage in relevant epistemic practices. Science literacy is not only about content knowledge, as previously discussed, but also about understanding of and engagement in scientific practices. Science game designers have to consider not only the content but also the mechanics of the game; the mechanics will have to convey the science concepts, relations and processes modelled by the game with respect to the learning content and objectives.

3.3.5 Context and Settings

Most of the studies reviewed were conducted in formal education settings (class-rooms), mainly in elementary, middle and high school, with very few exceptions focusing on preschoolers [e.g. Hsu et al. (2011)], higher education students (Hou 2015) or informal settings (playing at home) [e.g. Waddington and Fennewald (2018)]. In most cases though, the gameplay interventions in the classrooms were not part of the conventional school programme but rather an external intervention. The researchers cooperated with the teachers and examined the games and outcomes through experimental or semi-experimental conditions. In a number of cases, the researchers worked closely with education stakeholders such as teachers and school districts to develop games and curricula adapted to national curriculum objectives, and school needs and requirements (Ketelhut and Nelson 2010; Wallon et al. 2018). This is particularly important since learning outcomes seem to be affected by more than individual factors or the design of the game.

Considering issues of empowerment, equality and inclusion in science-related fields, the integration of science games in formal education classrooms seems to be particularly important. Repenning et al. (2015) described a critical issue when discussing the design of a middle-school curriculum for computer science education: self-selected student programmes such as after-school classes usually attract students already interested in the topic (i.e. computer science). Systemic integration of science-related games in schools would increase students' access to traditionally under-represented minority students and girls, similar to findings in (Voulgari and Yannakakis 2019). Only one study was found, though, to explicitly address learners with disabilities and appropriate game design: Marino et al. (2013) described and tested a game incorporating guidelines from the *Universal Design for Learning* (UDL) framework.

Students' interpretations of the game may vary beyond the game designers' intentions or expectations. The role of instruction and of the teacher is at that point important (Waddington and Fennewald 2018); the teachers, through reflection and discussions, can identify misconceptions and guide the students to view the game critically and consider alternative perspectives. Inversely, teachers' negative attitudes towards the game may also affect students' attitudes (Muehrer et al. 2012). Furthermore, students supported by material external to the game performed better in terms of the quality of their scientific argumentation (Wallon et al. 2018). In authentic classroom settings, other external factors such as slow internet connections, technical specifications of the computers and the conditions of the classroom (e.g. crowded, heat) can have an impact on the students' gameplay (Muehrer et al. 2012). The teachers, the quality of instruction and the surrounding conditions were, therefore, also found to factor in the learning outcomes.

3.4 Conclusions

In this chapter, we reviewed studies involving digital games and science learning. Our goal was to address not only the games as instructional tools, but also view them in a broader context involving cultural and societal practices. We observed a shift to higher order thinking skills and scientific practices such as inquiry, problem-solving and scientific reasoning, which is encouraging considering that such skills are important for the students to develop a critical and sceptical attitude in their lives. Even so, though, there is still great potential for research focused on the development of science literacy through social science-related games.

Although previous studies have described digital games as media that can trigger the interest for science and technology (Biles 2012; Bricker and Bell 2012; Mayo 2009), research and game development on this area is still limited. Further studies, for instance, could focus on the relation between the *science capital* of the students and their game preferences and propose game elements that can engage students with lower science capital scores.

Similarly, game studies and development could further consider the limited participation of girls in STEM-related fields (Dasgupta and Stout 2014) and the limited focus on games for children with intellectual, learning, sensory or motor disabilities (Beeston et al. 2018; Brown et al. 2010), and address these populations' requirements as well.

One of the main strengths of games is their potential as affective environments for science learning (Li and Tsai 2013). Fun, engagement, immersion and motivation in the studies reviewed have mainly been measured via surveys and self-reports. Research on the affective aspects of games based on biometric data and psychophysiological measures would provide more objective and valid data on the emotions and experience of the learners (Yannakakis and Martínez 2015).

Most of the studies reviewed focus on formal education settings. Games, though, are widely used in informal or non-formal learning settings, such as after-school programmes, science-fairs, FabLabs, Game Jams, or at home, supporting informal learning and the emergence of communities of practice spontaneously formed by even younger children and having a great educational potential (Arya et al. 2013; Squire and Patterson 2009; Williamson and Facer 2004). It seems that further research in informal and non-formal learning settings would yield valuable insights into the processes and factors involved. Research, though, in such settings presents challenges such as the lack of uniformity in learning objectives, and the varying attitudes and diversity of the participants (Honey and Hilton 2011, p. 78; Tisza et al. 2019).

Although this was not an extensive, empirical review of the literature, it did allow us to identify potential areas of interest for further research and design of digital games for science learning and scientific literacy. We tried to view the topic through a wider lens involving the game design, individual factors, as well as the social and cultural context considering the importance of media and digital literacy skills for children's education.

Acknowledgements The author wishes to thank Prof. Georgios Yannakakis for his collaboration, support and comments on this study. This work is supported by the "Learning science the fun and creative way: coding, making, and play as vehicles for informal science learning in the 21st century (COM n PLAY Science)" project, under the European Commission's Horizon 2020 SwafS-11-2017 Program (Project Number: 787476) and the "Learn to Machine Learn" (LearnML) project, under the Erasmus+ Strategic Partnership programme (Project Number: 2019-1-MT01-KA201-051220).

References

Archer, L., Dawson, E., DeWitt, J., Seakins, A., & Wong, B. (2015). "Science capital": A conceptual, methodological, and empirical argument for extending Bourdieusian notions of capital beyond the arts. *Journal of Research in Science Teaching, 52*(7), 922–948. https://doi.org/10.1002/tea.21227.

Arya, A., Chastine, J., Preston, J., & Fowler, A. (2013). An international study on learning and process choices in the global game jam. *International Journal of Game-Based Learning (IJGBL), 3*(4), 27–46. https://doi.org/10.4018/ijgbl.2013100103.

Ault, M., Craig-Hare, J., Frey, B., Ellis, J. D., & Bulgren, J. (2015). The effectiveness of reason racer, a game designed to engage middle school students in scientific argumentation. *Journal of Research on Technology in Education, 47*(1), 21–40. https://doi.org/10.1080/15391523.2015. 967542.

Beeston, J., Power, C., Cairns, P., & Barlet, M. (2018). *Characteristics and motivations of players with disabilities in digital games.* ArXiv:1805.11352 [Cs]. https://arxiv.org/abs/1805.11352.

Beier, M. E., Miller, L. M., & Wang, S. (2012). Science games and the development of scientific possible selves. *Cultural Studies of Science Education, 7*(4), 963–978. https://doi.org/10.1007/ s11422-012-9408-0.

Bergey, B. W., Ketelhut, D. J., Liang, S., Natarajan, U., & Karakus, M. (2015). Scientific inquiry self-efficacy and computer game self-efficacy as predictors and outcomes of middle school boys' and girls' performance in a science assessment in a virtual environment. *Journal of Science Education and Technology, 24*(5), 696–708. https://doi.org/10.1007/s10956-015-9558-4.

Biles, M. (2012). Leveraging insights from mainstream gameplay to inform STEM game design: Great idea, but what comes next? *Cultural Studies of Science Education, 7*(4), 903–908. https:// doi.org/10.1007/s11422-012-9453-8.

Bricker, L. A., & Bell, P. (2012). "GodMode is his video game name": Situating learning and identity in structures of social practice. *Cultural Studies of Science Education, 7*(4), 883–902. https://doi.org/10.1007/s11422-012-9410-6.

Brown, D. J., Standen, P., Evett, L., Battersby, S., & Shopland, N. (2010). Designing serious games for people with dual diagnosis: Learning disabilities and sensory impairments. In *Design and implementation of educational games: Theoretical and practical perspectives* (pp. 424–439). IGI Global. https://doi.org/10.4018/978-1-61520-781-7.ch027.

Cannady, M. A., Vincent-Ruz, P., Chung, J. M., & Schunn, C. D. (2019). Scientific sensemaking supports science content learning across disciplines and instructional contexts. *Contemporary Educational Psychology, 59,* 101802. https://doi.org/10.1016/j.cedpsych.2019.101802.

Chee, Y. S., & Tan, D.K.-C. (2012). Becoming chemists through game-based inquiry learning: The case of legends of Alkhimia. *Electronic Journal of E-Learning, 10*(2), 185–198.

Chen, C.-H., Wang, K.-C., & Lin, Y.-H. (2015). The comparison of solitary and collaborative modes of game-based learning on students' science learning and motivation. *Journal of Educational Technology & Society, 18*(2), 237–248. JSTOR.

Cheng, M. T., She, H. C., & Annetta, L. A. (2015). Game immersion experience: Its hierarchical structure and impact on game-based science learning. *Journal of Computer Assisted Learning, 31*(3), 232–253. https://doi.org/10.1111/jcal.12066.

Cheng, M.-T., Chen, J.-H., Chu, S.-J., & Chen, S.-Y. (2015). The use of serious games in science education: A review of selected empirical research from 2002 to 2013. *Journal of Computers in Education, 2*(3), 353–375. https://doi.org/10.1007/s40692-015-0039-9.

Cheng, M.-T., Su, T., Huang, W.-Y., & Chen, J.-H. (2014). An educational game for learning human immunology: What do students learn and how do they perceive? *British Journal of Educational Technology, 45*(5), 820–833. https://doi.org/10.1111/bjet.12098.

Clark, D. B., Sengupta, P., Brady, C. E., Martinez-Garza, M. M., & Killingsworth, S. S. (2015). Disciplinary integration of digital games for science learning. *International Journal of STEM Education, 2*(1), 2. https://doi.org/10.1186/s40594-014-0014-4.

Clark, D., Nelson, B., Sengupta, P., & D'Angelo, C. (2009). Rethinking science learning through digital games and simulations: Genres, examples, and evidence. *Learning science: Computer games, simulations, and education workshop sponsored by the National Academy of Sciences, Washington, DC.*

Cordova, D. I., & Lepper, M. R. (1996). Intrinsic motivation and the process of learning: Beneficial effects of contextualization, personalization, and choice. *Journal of Educational Psychology, 88,* 715–730.

Curtis, V. (2015). Motivation to participate in an online citizen science game: A study of Foldit. *Science Communication, 37*(6), 723–746. https://doi.org/10.1177/1075547015609322.

Dasgupta, N., & Stout, J. G. (2014). Girls and women in science, technology, engineering, and mathematics: STEMing the tide and broadening participation in STEM careers. *Policy Insights from the Behavioral and Brain Sciences, 1*(1), 21–29. https://doi.org/10.1177/237273221454 9471.

Dippel, A., & Fizek, S. (2019). Laborious playgrounds: Citizen science games as new modes of work/play in the digital. In R. Glas, S. Lammes, M. de Lange, J. Raessens, & I. de Vries (Eds.), *The playful citizen: Civic engagement in a mediatized culture.* Amsterdam University Press. https://www.jstor.org/stable/10.2307/j.ctvcmxpds.

Egenfeldt-Nielsen, S. (2007). Third generation educational use of computer games. *Journal of Educational Multimedia and Hypermedia, 16*(3), 263–281.

Fraser, J., Shane-Simpson, C., & Asbell-Clarke, J. (2014). Youth science identity, science learning, and gaming experiences. *Computers in Human Behavior, 41,* 523–532. https://doi.org/10.1016/j.chb.2014.09.048.

Gaydos, M. J., & Squire, K. D. (2012). Role playing games for scientific citizenship. *Cultural Studies of Science Education, 7*(4), 821–844. https://doi.org/10.1007/s11422-012-9414-2.

Halpern, D. F., Millis, K., Graesser, A. C., Butler, H., Forsyth, C., & Cai, Z. (2012). Operation ARA: A computerized learning game that teaches critical thinking and scientific reasoning. *Thinking Skills and Creativity, 7*(2), 93–100. https://doi.org/10.1016/j.tsc.2012.03.006.

Hidi, S. (1990). Interest and its contribution as a mental resource for learning. *Review of Educational Research, 60*(4), 549–571. https://doi.org/10.3102/00346543060004549.

Honey, M., & Hilton, M. L. (2011). *Learning science through computer games and simulations.* National Academies Press. https://doi.org/10.17226/13078.

Hou, H.-T. (2015). Integrating cluster and sequential analysis to explore learners' flow and behavioral patterns in a simulation game with situated-learning context for science courses: A video-based process exploration. *Computers in Human Behavior, 48,* 424–435. https://doi.org/10.1016/j.chb.2015.02.010.

Hsu, C.-Y., Tsai, C.-C., & Liang, J.-C. (2011). Facilitating preschoolers' scientific knowledge construction via computer games regarding light and shadow: The effect of the prediction-observation-explanation (POE) strategy. *Journal of Science Education and Technology, 20*(5), 482–493. https://doi.org/10.1007/s10956-011-9298-z.

Israel, M., Wang, S., & Marino, M. T. (2016). A multilevel analysis of diverse learners playing life science video games: Interactions between game content, learning disability status, reading proficiency, and gender. *Journal of Research in Science Teaching, 53*(2), 324–345. https://doi.org/10.1002/tea.21273.

Kafai, Y. B. (2006). Playing and making games for learning: Instructionist and constructionist perspectives for game studies. *Games and Culture, 1,* 36–40. https://doi.org/10.1177/155541200 5281767.

Ke, F. (2009). A qualitative meta-analysis of computer games as learning tools. In R. E. Ferdig (Ed.), *Handbook of research on effective electronic gaming in education (3 Volumes)* (Vol. 1, pp. 1–32). Information Science Reference. https://www.igi-global.com/reference/details.asp?ID=7960.

Ketelhut, D. J., & Nelson, B. C. (2010). Designing for real-world scientific inquiry in virtual environments. *Educational Research, 52*(2), 151–167. https://doi.org/10.1080/00131881.2010.482741.

Ketelhut, D. J., Nelson, B. C., Clarke, J., & Dede, C. (2010). A multi-user virtual environment for building and assessing higher order inquiry skills in science. *British Journal of Educational Technology, 41*(1), 56–68. https://doi.org/10.1111/j.1467-8535.2009.01036.x.

Koltay, T. (2011). The media and the literacies: Media literacy, information literacy, digital literacy. *Media, Culture & Society, 33*(2), 211–221. https://doi.org/10.1177/0163443710393382.

Koops, M., & Hoevenaar, M. (2013). Conceptual change during a serious game: Using a lemniscate model to compare strategies in a physics game. *Simulation & Gaming, 44*(4), 544–561. https://doi.org/10.1177/1046878112459261.

Lee, J. K., & Probert, J. (2010). Civilization III and whole-class play in high school social studies. *Journal of Social Studies Research, 34*(1), 1–28.

Lester, J. C., Spires, H. A., Nietfeld, J. L., Minogue, J., Mott, B. W., & Lobene, E. V. (2014). Designing game-based learning environments for elementary science education: A narrative-centered learning perspective. *Information Sciences, 264,* 4–18. https://doi.org/10.1016/j.ins.2013.09.005.

Li, M.-C., & Tsai, C.-C. (2013). Game-based learning in science education: A review of relevant research. *Journal of Science Education and Technology, 22*(6), 877–898. https://doi.org/10.1007/s10956-013-9436-x.

Liu, M., Rosenblum, J. A., Horton, L., & Kang, J. (2014). Designing science learning with game-based approaches. *Computers in the Schools, 31,* 84–102. https://doi.org/10.1080/07380569.2014.879776.

Magnussen, R., Hansen, S. D., Planke, T., & Sherson, J. F. (2014). Games as a platform for student participation in authentic scientific research. *Electronic Journal of E-Learning (EJEL), 2*(3), 259–270.

Marino, M. T., & Hayes, M. T. (2012). Promoting inclusive education, civic scientific literacy, and global citizenship with videogames. *Cultural Studies of Science Education, 7*(4), 945–954. https://doi.org/10.1007/s11422-012-9429-8.

Marino, M. T., Israel, M., Beecher, C. C., & Basham, J. D. (2013). Students' and teachers' perceptions of using video games to enhance science instruction. *Journal of Science Education and Technology, 22*(5), 667–680. https://doi.org/10.1007/s10956-012-9421-9.

Markus, H., & Nurius, P. (1986). Possible selves. *American Psychologist, 41*(9), 954–969.

Martinez-Garza, M., Clark, D. B., & Nelson, B. C. (2013). Digital games and the US National Research Council's science proficiency goals. *Studies in Science Education, 49*(2), 170–208. https://doi.org/10.1080/03057267.2013.839372.

Mayo, M. J. (2009). Video games: A route to large-scale STEM education? *Science, 323*(5910), 79–82. https://doi.org/10.1126/science.1166900.

Muehrer, R., Jenson, J., Friedberg, J., & Husain, N. (2012). Challenges and opportunities: Using a science-based video game in secondary school settings. *Cultural Studies of Science Education, 7*(4), 783–805. https://doi.org/10.1007/s11422-012-9409-z.

National Research Council. (2011). Learning science through computer games and simulations. Committee on science learning: Computer games, simulations, and education. In M. A. Honey & M. L. Hilton (Eds.), *Board on science education, division of behavioral and social sciences and education.* Washington, DC: The National Academies Press.

OECD. (2012). *Students, computers and learning: Making the connection.* PISA, OECD Publishing. https://dx.doi.org/10.1787/9789264239555-en.

O'Neil, H. F., Wainess, R., & Baker, E. L. (2005). Classification of learning outcomes: Evidence from the computer games literature. *Curriculum Journal, 16*(4), 455–474. https://doi.org/10.1080/09585170500384529.

Ponti, M., Hillman, T., Kullenberg, C., & Kasperowski, D. (2018). Getting it right or being top rank: Games in citizen science. *Citizen Science: Theory and Practice, 3*(1), 1. https://doi.org/10.5334/cstp.101.

Repenning, A., Webb, D. C., Koh, K. H., Nickerson, H., Miller, S. B., Brand, C., et al. (2015). Scalable game design: A strategy to bring systemic computer science education to schools through game design and simulation creation. *Transactions on Computing Education, 15*(2), 11:1–11:31. https://doi.org/10.1145/2700517.

Ryan, R. M., Rigby, C. S., & Przybylski, A. (2006). The motivational pull of video games: A self-determination theory approach. *Motivation and Emotion, 30*(4), 347–363.

Sáez-López, J.-M., Miller, J., Vázquez-Cano, E., & Domínguez-Garrido, M.-C. (2015). Exploring application, attitudes and integration of video games: MinecraftEdu in middle school. *Educational Technology & Society, 18*(3), 114–128.

Schifter, C. C., Ketelhut, D. J., & Nelson, B. C. (2012). Presence and middle school students' participation in a virtual game environment to assess science inquiry. *Journal of Educational Technology & Society, 15*(1), 53–63. JSTOR.

Shaffer, D. W. (2006). Epistemic frames for epistemic games. *Computers & Education, 46*(3), 223–234. https://doi.org/10.1016/j.compedu.2005.11.003.

Squire, K., & Patterson, N. (2009). *Games and simulations in informal science education* [Paper commissioned for the National Research Council Workshop on Gaming and Simulations]. The National Academies of Sciences, Engineering, Medicine. https://sites.nationalacademies.org/DBASSE/BOSE/DBASSE_071087.

Sun, H., & Gao, Y. (2016). Impact of an active educational video game on children's motivation, science knowledge, and physical activity. *Journal of Sport and Health Science, 5*(2), 239–245. https://doi.org/10.1016/j.jshs.2014.12.004.

Taub, M., Azevedo, R., Bradbury, A. E., Millar, G. C., & Lester, J. (2018). Using sequence mining to reveal the efficiency in scientific reasoning during STEM learning with a game-based learning environment. *Learning and Instruction, 54,* 93–103. https://doi.org/10.1016/j.learninstruc.2017.08.005.

Tisza, G., Papavlasopoulou, S., Christidou, D., Voulgari, I., Iivari, N., Giannakos, M. N., et al. (2019). The role of age and gender on implementing informal and non-formal science learning activities for children. In *Proceedings of the FabLearn Europe 2019 Conference* (pp. 10:1–10:9). https://doi.org/10.1145/3335055.3335065.

VanFossen, P. P. J., Friedman, A., & Hartshorne, R. (2009). The role of MMORPGs in social studies education. In R. E. Ferdig (Ed.), *Handbook of research on effective electronic gaming in education* (pp. 235–250). IGI Global. https://www.igi-global.com/chapter/handbook-research-effective-electronic-gaming/20089/.

Voulgari, I., & Yannakakis, G. N. (2019). Digital games in non-formal and informal learning practices for science learning: A case study. In A. Liapis, G. N. Yannakakis, M. Gentile, & M. Ninaus (Eds.), *Games and learning alliance* (pp. 540–549). Springer International Publishing. https://doi.org/10.1007/978-3-030-34350-7_52.

Waddington, D. I., & Fennewald, T. (2018). Grim FATE: Learning about systems thinking in an in-depth climate change simulation. *Simulation & Gaming, 49*(2), 168–194. https://doi.org/10.1177/1046878117753498.

Wallon, R. C., Jasti, C., Lauren, H. Z. G., & Hug, B. (2018). Implementation of a curriculum-integrated computer game for introducing scientific argumentation. *Journal of Science Education and Technology, 27*(3), 236–247. https://doi.org/10.1007/s10956-017-9720-2.

Williamson, B., & Facer, K. (2004). More than 'just a game': The implications for schools of children's computer games communities. *Education, Communication & Information, 4,* 255–270. https://doi.org/10.1080/14636310412331304708.

Yannakakis, G. N., & Martínez, H. P. (2015). Ratings are overrated! *Frontiers in ICT, 2.* https://doi.org/10.3389/fict.2015.00013.

Iro is a postdoctoral researcher at the Institute of Digital Games, University of Malta and teaching staff at the Department of Early Childhood Education, National and Kapodistrian University of Athens. She is teaching undergraduate and postgraduate courses on Digital Games and Virtual Worlds, and Learning Technologies. Her research focuses on game-based learning, game studies, and digital literacy. She has organised several workshops relevant to game-based learning, Information and Communication Technologies in Education, and Digital Storytelling in local and international venues. She has worked on several Nationally and EU-funded research projects on the design, implementation and assessment of Learning Technologies in teaching and learning.

Chapter 4
Web-Based Learning in Computer Science: Insights into Progress and Problems of Learners in MOOCs

Johannes Krugel and Peter Hubwieser

Abstract Web-based resources and massive open online courses (MOOCs) are promising forms of non-formal and technology-enhanced learning. Advantages are the flexibility regarding location and time and the possibilities for self-regulated learning. Furthermore, web technologies have considerably evolved over the past years, enabling complex interactive exercises and communication among the learners. However, online learning also has its challenges regarding, e.g., the motivation and low-completion rates in MOOCs. Following a design-based research methodology, we designed, developed, and evaluated a MOOC for the introduction of object-oriented programming. In three course runs, we collected extensive textual feedback from the participants which we analyzed inductive qualitative content analysis (QCA) by Mayring. We complement this with quantitative analyses regarding the performance of the learners in the course. The results give insights into the progress, preferences, and problems of learners in MOOCs. We furthermore used these results as a basis for adapting the course in the following iterations of our design-based research and observed a significant increase in the course completion rate.

4.1 Introduction

Web-based learning is a form of technology-enhanced learning with the advantage that the technical barriers are very low, for the learners as well as for the creators. Massive open online courses (MOOCs) combine several web-based learning activities in the form of a course. MOOCs became an educational buzzword in 2012 and have enjoyed wide media coverage in the popular press. In contrast to traditional ways of teaching, where the size of participants is restricted, MOOCs have to be easily scalable for large numbers of participants. They are usually free for everybody and

J. Krugel (✉) · P. Hubwieser
School of Education, Technical University of Munich, Munich, Germany
e-mail: krugel@tum.de

P. Hubwieser
e-mail: peter.hubwieser@tum.de

© Springer Nature Singapore Pte Ltd. 2020
M. Giannakos (ed.), *Non-Formal and Informal Science Learning in the ICT Era*, Lecture Notes in Educational Technology,
https://doi.org/10.1007/978-981-15-6747-6_4

can be used for formal as well as non-formal learning. According to the *e-learning hype curve*, MOOCs are already through the "Trough of Disillusionment" and nearly on the "slope of enlightenment" (Hicken 2018). Our focus in this paper is MOOCs for the introduction of programming, especially object-oriented programming.

4.1.1   Our MOOC

Computer science (CS) education in school is varying strongly in many countries. In Germany, for example, the implementation of CS education at school is very diverse, unregulated, and inconsistent in many states. In consequence, the prerequisite knowledge of freshmen at universities is very inhomogeneous (Hubwieser et al. 2015). As students cannot be expected to be present at university before lecturing starts, MOOCs (massive open online courses) seem to represent potential solutions to compensate or reduce these differences. This was the initial motivation to develop our MOOC called "LOOP: Learning Object-Oriented Programming".

The initial primary target group of the course is prospective students of science or engineering that are due to attend CS lessons in their first terms. However, since the course is available online and free for everybody, the target group now is a worldwide audience (speaking German) and with a more diverse background.

As learning to program is a substantial cognitive challenge (Hubwieser 2008), MOOCs run in danger to overstrain the students, frustrating them already before their studies. To meet this challenge, we carefully designed LOOP, starting with a gentle introduction to computational thinking (Wing 2006). The course is based on the *strictly objects first* approach (Gries 2008), introducing the concepts *object*, *attribute*, and *method* just before *class* and before any programming activity. This helps to avoid excessive cognitive load following when it comes to actually write programs in an object-oriented programming language (in our course Java).

The course includes various interactive exercises to enable the learners to experiment with the presented concepts. Furthermore, we implemented programming exercises with constructive feedback for the learners using a web-based integrated development environment and additionally an automatic grading system.

A detailed description of the rationale behind the course design and the course curriculum was published in (Krugel and Hubwieser 2018). The results of a pilot run of the course are presented in (Krugel and Hubwieser 2017).

4.1.2   Research Questions and Research Design

In this paper, we examine the learners' perspectives to gain insights into the progress, problems, and preferences of learners in MOOCs. In particular, we aim to answer the following research questions:

1. What are learners' preferences in self-regulated introductory programming MOOCs?
2. What are the main challenges learners face in self-regulated introductory programming MOOCs?
3. What are the main reasons for not finishing introductory programming MOOCs?

The answers to those research questions are going to help to design self-learning courses and improve existing courses. Following a design-based research methodology, we start with a literature review, implement a prototypical course, assess the learning, and proceed in iterations (Design-Based Research Collective 2003; Plomp 2007).

We collect both quantitative and qualitative data, utilizing a mixed-methods approach for the data analysis; this approach makes it possible to gain in-breadth and in-depth understanding while balancing the weaknesses inherent in the use of each individual approach. To understand the learner's perspectives, we collected and analyzed feedback of the course participants. We applied an inductive qualitative content analysis to categorize the responses. In terms of learning progress, we pursue quantitative methods and conduced a hierarchical cluster analysis of participants' scores in the assignments. We use the results of the analysis to adapt the teaching and perform more iterations, observing the changes on the learners' side.

This paper is structured as follows. As a background, we describe the design-based research methodology, the foundations for the course design, and further related literature. Then we present the course design and its curriculum. The main part of this paper is the description of our data analysis methodology and the results. We conclude with a discussion of the limitations and an outlook.

4.2 Background and Related Work

In the following, we briefly describe the background of the research methodology, related MOOCs for introductory CS, research on MOOC design and drop-out behavior.

4.2.1 Design-Based Research (DBR)

DBR is an empirical approach where the application and investigation of theories is closely connected with the design and evaluation of learning interventions (Design-Based Research Collective 2003). There is not a unique definition of the term but many authors agree that DBR encompasses a preliminary research, a prototyping phase, and an assessment phase (Plomp 2007).

According to Reeves (2006), DBR consists of four phases: the analysis of a practical problem, the development of a solution, testing and refinement of the solution in

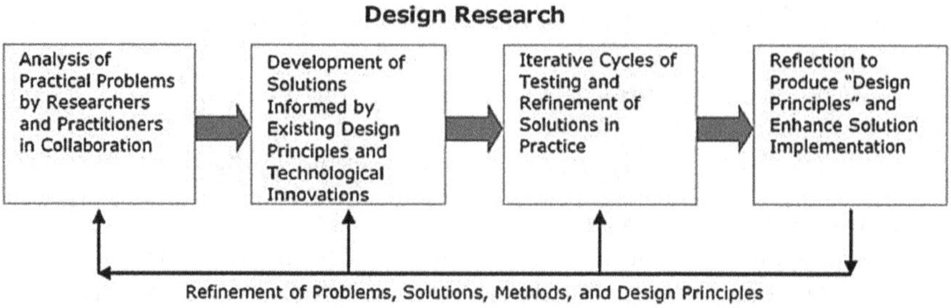

Design Research

Fig. 4.1 Phases of design-based research (Reeves 2006)

practice, and a reflection to produce design principles, see Fig. 4.1. A main characteristic of DBR is the systematic and iterative cycle of design, exploration, and redesign (Design-Based Research Collective 2003). The evaluation is often carried out using a mixed-methods approach (Anderson and Shattuck 2012). Using mixed-methods for the evaluation makes it possible to gain in-breadth and in-depth insights while balancing the weaknesses inherent in the use of each individual approach.

DBR is nowadays widely used in different contexts of educational research, also for learning programming and algorithmic thinking (Geldreich et al. 2019; Papavlasopoulou et al. 2019).

4.2.2 Research on MOOCs

There is an uncountable number of online courses for learning the basics of computer science. Some examples of massive open online courses (MOOCs) and small private online courses (SPOCs) that explicitly cover computational thinking or object-oriented programming (OOP) and were published in the scientific literature are described by Liyanagunawardena et al. (2014), Piccioni et al. (2014), Falkner et al. (2016), Alario-Hoyos et al. (2016), Kurhila and Vihavainen (2015), Vihavainen et al. (2012), Fitzpatrick et al. (2017), and Bajwa et al. (2019).

Several studies investigate the reasons of the learners for enrolling in MOOCs. Crues et al. (2018) analyzed the responses to open-ended questions in five MOOCs; they used methods from natural language processing (Latent Dirichlet Allocation) and identified a set of 26 "topics" as reasons to enroll in a MOOC.

Luik et al. (2019a, b) developed and tested an instrument to measure the motivational factors for enrolling in a programming MOOC.

Zheng et al. (2015) carried out interviews with participants of several MOOCs and analyzed them using grounded theory; among others, they investigated reasons for enrollment and how students learn in a MOOC.

In general, MOOCs are known to have a rather high dropout (Delgado Kloos et al. 2014; Garcia et al. 2014; Piccioni et al. 2014) with completion rates usually in the range of 5–10%.

Zheng et al. (2015) also analyzed the interviews regarding the reasons for not completing a MOOC and identified eight categories: *High Workload, Challenging Course Content, Lack of Time, Lack of Pressure, No Sense of Community or Awareness of Others, Social Influence, Lengthy Course Start-Up, Learning on Demand*.

In another study, Eriksson et al. (2017) carried out interviews with participants of two MOOCs; the main result regarding the drop-out in MOOCs is that "Time is the bottleneck".

To assess the learner's perspectives in MOOCs with a special focus on active learning, Topali et al. (2019) analyzed questionnaires, forum posts, and logs of the learning platform; they discuss challenges of the learners and reasons for not completing the course.

Luik et al. (2019a, b) investigate the connection of drop-out with demographic factors of the participants in three programming MOOCs. There are, furthermore, many approaches to predict the drop-out probability of the learners based on the data available in such online courses, see e.g., (Moreno-Marcos et al. 2020).

4.3 Course Design

According to the DBR methodology by Reeves (2006), we started by analyzing the needs of the learners and developing a solution based on existing design principles (phase 1 and phase 2). The course consists of a series of short videos, followed by quizzes, interactive exercises, and a final exam. The communication of the learners and with the course team takes place in a discussion board. In the following, we describe those course elements, and present the details of the course design.

4.3.1 Videos

All topics of the course are presented in short videos with an average length of 5 min. The videos were produced based on the suggestions of Guo et al. (2014) and similar to the suggestions by Alonso-Ramos et al. (2016) published shortly after our recording.

Each of the 24 videos begins with a short advance organizer to help the learners focus on the relevant aspects. This is augmented with the talking head of the respective instructor (using chroma key compositing) facilitating the learners to establish a personal and emotional connection (Alonso-Ramos et al. 2016).

For the actual content of the videos, we decided to use a combination of slides and tablet drawing. The background of the video consists of presentation slides and the instructor uses a tablet to draw and develop additional aspects or to highlight important part of the slides. All slides are provided for download and we additionally added audio transcripts for the videos. By such video, audio, and textual representations, several senses are addressed simultaneously, making the content accessible to learners with different learning preferences or impairments.

4.3.2 Quizzes

After each video, the course contains quizzes as formative assessment. The main purpose is to provide the learners with direct and instant feedback on the learning progress. The quizzes use the standard assessment types offered by the MOOC platform, e.g., single- /multiple-choice questions, drop-down lists, drag-and-drop problems, or text input problems. Depending on the answer, the learner gets a positive feedback or, otherwise, for example, hints which previous parts of the course to repeat in more detail.

4.3.3 Interactive Exercises

The videos introduce new concepts to the learners and the quizzes test the progress, which is, however, in general not sufficient to acquire practical competencies (Alario-Hoyos et al. 2016). Following a rather constructivist approach, we intend to let the learners experiment and interact with the concepts directly. Considering that, we include interactive exercises or programming task for all learning steps throughout the course. Special care was devoted to the selection and development of those interactive exercises to enable the learners to experiment and interact directly with the presented concepts. It can be a major obstacle for potential participants having to install special software (Liyanagunawardena et al. 2014; Piccioni et al. 2014), which is especially problematic in an online setting without a teacher who could help in person. We, therefore, decided to use only purely web-based tools. There are already many web-based tools for fostering computational thinking and learning OOP concepts available in the internet. We selected the in our opinion most suitable tools to support the intended learning goals. Where necessary we adapted or extended them to meet our needs. All tools are integrated seamlessly into the learning platform resulting in a smooth user experience.

4.3.4 Programming Exercises

While in several introductory CS MOOCs the learners have to install an integrated development environment (IDE) for writing their first computer programs, we decided to rely on web-based tool also for this purpose [like (Piccioni et al. 2014)]. We chose to use *Codeboard* (Estler and Nordio) among several alternatives (Derval et al. 2015; Skoric et al. 2016; Staubitz et al. 2016) because of the usability and seamless integration into the *edX* platform using the Learning Tools Interoperability (LTI) standard.

The programming assignments are graded automatically and the main purpose is to provide instant feedback to the learner. We, therefore, implemented tests for each assignment that make heavy use of the Java reflection functionality. While standard unit tests would fail with a compile error if, e.g., an attribute is missing or spelled differently. Reflection makes it possible to determine for a learner's submission if, e.g., all attributes and methods are defined with the correct names, types, and parameters. Writing the tests requires more effort than for standard unit tests but can give more detailed feedback for the learners in case of mistakes.

Additionally, we integrated the automatic grading and feedback system *JACK* (Striewe and Goedicke 2013) using the external grader interface of the *edX* platform. Apart from static and dynamic tests, *JACK* also offers the generation and comparison of traces and visualization of object structures; however, we do not use this extended functionality yet.

4.3.5   Discussion Board

The course also provides a discussion forum which is divided into several discussion boards. The communication among the learners and with the instructors is supposed to take place entirely in the discussion forum. Besides the default board for general and organizational issues, we created one separate discussion board for each course chapter (called chapter discussion board in the following). The idea is to organize the learners' discussions, such that it is easier for the learners to keep an overview. In those discussion boards, the learners can ask questions and answer the questions of others. The course team also tracks the discussions and can intervene when problems cannot be solved timely by the learning community itself. It is also possible to vote for popular topics but we did not mention this possibility and nearly nobody used it.

Additionally, we included separate discussion boards for six specific exercises in which the learners are supposed to upload their solutions to the forum (those boards are called exercise discussion boards in the following). In the first of those exercises, the learners are encouraged to introduce themselves to the community giving some information about their age, where they live, etc. In the exercise discussion boards, the learners are also prompted to give feedback to the contributions of others.

4.3.6   Final Exam

The course contains a final exam at the end that covers topics of the whole course. The suggested time to work on the exam is 60 min; the time limit is, however, not ensured by technical limitations. The final exam as well as the graded exercises count for a final score. A minimal score of 50% is necessary to successfully pass the course.

4.3.7 Course Syllabus

Computational thinking (CT) as introduced by Wing (2006) is a universal personal ability that can be used in many disciplines. Since the target group of our course comes from various different fields of study, we incorporated CT as integral part of the course. CT is on the one hand intended to facilitate learning programming and on the other hand a sustainable competency that can be used also outside of our course.

As pointed out in (Hubwieser 2008), there is a fundamental didactical dilemma in teaching OOP: On the one hand, modern teaching approaches postulate to teach in a "real life" context (Cooper and Cunningham 2010), i. e., to pose authentic problems to the learners. Therefore, it seems advisable to start with interesting, sufficiently complex tasks that convince the learners that the concepts they have to learn are helpful in their professional life. However, if we start with such problems, we might ask too much from the learners, because they will have to learn an enormous amount of new, partly very difficult concepts at once (Hubwieser 2008).

Following a *strictly objects first* approach (Gries 2008) and similar to the design of the school subject and an introductory university lecture, we solved this problem by distributing the learning objectives over the parts of the course that precede the "serious" programming part. This avoids to confront the learners with too many unknown concepts when they have to write their first program. Among others, we suggest to the learners to look at an object as a state machine (Hubwieser 2008). In order to realize this in a learner-oriented way, the learners need to be able to understand a simulation program of a typical state machine, e.g, a traffic light system.

Concerning the choice of the examples, we set the emphasis on the relevance for the everyday life, which leads for instance to banking or domestic appliances.

LOOP consists of the following five chapters and 16 sections:

1. Object-oriented modeling

   - 1.1 Objects
   - 1.2 Classes
   - 1.3 Methods and parameters
   - 1.4 Associations
   - 1.5 States of objects.

2. Algorithms

   - 2.1 Concept of algorithm
   - 2.2 Structure of algorithms.

3. Classes in programming languages

- 3.1 Class definition
- 3.2 Methods
- 3.3 Creation of objects.

4. Object-oriented programming

   - 4.1 Implementing algorithms
   - 4.2 Arrays.

5. Associations and references

   - 5.1 Aggregation and references
   - 5.2 Managing references
   - 5.3 Communication of objects
   - 5.4 Sequence charts.

A detailed description of the course design and syllabus was published in (Krugel and Hubwieser 2018).

4.4   Methodology and Results

We follow a design-based research methodology and in this work we focus on the assessment of our learning intervention [phase 3 and phase 4 according to Reeves (2006)]. Collecting both quantitative and qualitative data in this case study, we utilize a mixed-methods approach for the data analysis to gain in-breadth and in-depth understanding while balancing the weaknesses inherent in the use of each individual approach. To understand the learner's perspectives, we collected and analyzed feedback of the course participants. For the analysis, we applied an inductive qualitative content analysis according to Mayring (2000). In terms of learning progress, we pursue quantitative methods and conduced a hierarchical cluster analysis of participants' scores in the assignments. This calculation is performed using the programming language R.

In the following, we first inform about the course implementation, our data collection and give some data on the course participants. We then describe the data analysis with its results. We performed three DBR cycles so far and describe the adaptations of the course, which were based on the results of our analyses.

4.4.1 Course Implementation

We prepared the online course on the learning platform *edX*[1] and offered it three times in the summer holidays of 2016, 2017, and 2018. We will refer to a course run by its year in the following exposition.

Course run 2016 was offered as a small private online course (SPOC) on *edX Edge* and was announced internally at our university as a preparation course for CS basics. The course runs 2017 and 2018 were offered publicly as MOOCs on *edX*. For organizational reasons, 2017 actually consisted of two identical and directly consecutive course runs which we treat as one course run in the following. The course runs 2017 and 2018 were included in the global *edX* course catalog in the category *Computer sciences courses* and available worldwide. Our university announced the courses on its official Facebook page and informed all students in CS-related subjects about the course offerings.

The intended effort of the students is 5 h per week. Everyone was free to participate in the courses without any formal or content-specific prerequisites. The course takes into account different learning preferences and impairments by providing the learning content as visual (videos, graphics), textual, and audio presentations. The only requirement was German language proficiency since the course was offered on German.

Participation was voluntary in all course runs and did not count toward a grade. In 2016, we issued informal certificates for successful participation (=obtaining at least 50% of the possible points in at least 12 of 16 course units). In 2017 and 2018, *edX* issued verified certificates for successful participation, which, however, had to be paid (49 $).

Each course run took five weeks (one week for each chapter) and the targeted workload of the learners was 5–10 h per week. The communication among the learners and with the instructors took place entirely in the discussion forum. The main task of the instructor during the conduction of the course was to monitor the forum and to react accordingly, e.g, answer questions or fix problems with the grading system.

4.4.2 Data Collection

We integrated an introductory online questionnaire into the course (called "course start survey" in the following), in which we asked the participants about their age, gender, major, and previous programming experience.

In a concluding questionnaire at the end of the course (called "course end survey" in the following), we asked for positive and negative textual feedback regarding the course; it consists of two text field with the following questions (translated from German):

---

[1] https://www.edx.org.

1. Positive aspects: *Which aspects of the course did you like?*
2. Negative aspects: *Which problems did you encounter during the course? Which aspects of the course did you not like? Which suggestions for improvements do you have?*

The course end survey also asked for the approximate weekly workload that the learners spent on the course.

To get an even more detailed picture in the course runs 2017 and 2018, we additionally requested the participants' perspectives already during the course. We, therefore, included the same questions 1 and 2 directly after each chapter, for organizational reasons with one combined text input field for both questions together (this is called "chapter feedback survey" in the following). We can, therefore, react more specifically regarding the current chapter and also earlier (even immediately during the course run). This chapter feedback also enables us to get responses by those who do not make it to the end of the course and to assess what they think before actually dropping out.

We are especially interested in why some learners successfully finish the course and others do not. We, therefore, wrote an individualized mail to every course participant who did not complete the course (course runs 2017 and 2018). In the mail, we simply asked for the reason for not finishing the course (this is called "drop-out survey" in the following, even though not everybody not finishing a course is necessarily an actual drop-out).

Further, all responses to the quizzes and other tasks, as well as all programs submitted by the participants during the online course were collected. From the postings in the discussion forum even more qualitative data was obtained.

4.4.3   Participants

The three course runs of LOOP in 2016, 2017, and 2018 attracted $87 + 2390 + 2520 = 4997$ registrations. The following numbers are always the total of all course runs in which we collected the respective data if not stated otherwise.

For the course start survey, we received $80 + 494 + 1039 = 1613$ responses (female: 463, male: 992, diverse: 7, no answer: 151) with a diverse study background (numerous different majors, including Computer Science, Management, Engineering, Mathematics and many more). The participants come mainly from Germany, but in total from more than 70 countries. The average age was 23.5 and 315 participants were less than 20 years old. Regarding programming, 459 participants had no experience, 673 had basic knowledge, and 300 participants had already written a "bigger" program of at least 100 lines of code (no answer: 181 participants).

4.4.4 Learners' Feedback

To assess the learners' perspectives, we analyzed all their free-text responses. This reveals how the participants perceived and dealt with the online course. One goal of this analysis is to gain generalizable insights into the preferences and problems of learners in programming MOOCs. Another goal is to improve our course accordingly aiming at a higher satisfaction of the learners and a higher success rate.

Data Analysis

We chose to analyze the learners' utterances of the course end survey, the chapter feedback and the drop-out survey using qualitative content analysis following the methodology of Mayring (2000). This allows to categorize the statements which help to afterwards group and inspect similar statements. This will help to understand the learners' perspectives and to prioritize aspects of the course that can be improved. We are interested in the following three aspects which, therefore, guided the categorization:

1. Positive aspects of the course (course end survey and chapter feedback)
2. Negative aspects of the course (course end survey and chapter feedback)
3. Reasons for not finishing the course (drop-out survey).

We abstracted each learners' statement such that its wording is independent of our particular course: a learner wrote for example "*The drawing exercise took a lot of effort.*" and we abstracted as "*high effort*". We furthermore rephrased all abstractions as statements about the current state of the course: a learner wrote for example the suggestion "*It would be better to include more examples*" and we abstracted this as "*too little examples*". Since each learner's statement could consist of several sentences or even paragraphs, the rephrased versions could consist of multiple abstractions. We then inductively categorized all abstractions using several iterations over the data. In those iterations, we introduced new categories, refined, and merged existing categories. We kept track of the category systems maintaining a coding manual with descriptions of the categories. Whenever the category system changed, we performed another iteration over the corresponding data until reaching a stable state.

The result of this analysis is a classification of the learners' free-text responses of all course runs into the categories of the three supercategories while each response can fall into several categories.

Results for the Course End Survey

In the course end survey we received $13 + 102 + 191 = 306$ answers. The learners reported their views about the videos and exercises, the level of difficulty, technical and organizational issues. They furthermore described their individual progress or problems and proposed specific changes, among others. Due to the low number of responses in 2016, we did not include these responses in our further analysis.

Applying the inductive qualitative content analysis on the course end survey and the chapter feedback, we encountered 15 categories for the positive aspects and 45

Table 4.1 Categorization of negative aspects mentioned by the participants in the course end survey

2017 (%) (102)	2018 (%) (191)	Positive aspect	Details
6.9	3.7	Automatic feedback	For quizzes and programming tasks
2.0	0.0	Contents	Selection of course contents
11.8	9.4	Course structure	Variability and alignment of the elements
5.9	3.7	Discussion board	Help by other learners and the course team
6.9	4.2	Examples	Real-world connection, clarity
31.4	30.4	Exercises	Interactivity
5.9	10.5	Explanations	Clarity, understandability
5.9	3.1	External tools	Integration of external tools into the course
1.0	0.0	Final exam	Tasks and alignment of the exam
4.9	3.1	Flexibility	Regarding time and location
3.9	6.8	Handouts	Preparation, layout, downloadable
11.8	9.9	Level of difficulty	Not too easy, not too hard
14.7	15.7	Programming tasks	Variability and alignment of the elements
6.9	5.2	Quizzes	Help by other learners and the course team
16.7	18.8	Videos	Video style, availability

categories for the negative aspects taking all three course runs together. The results for the course end survey are displayed in Tables 4.1 and 4.2. The tables further contain for each category a description and the frequency within responses of the three course runs. Since this is the result from a *qualitative* analysis, we sorted the categories alphabetically and not by frequencies.

There are some categories listed with 0 occurrences. They stem from the chapter feedback because we used the same category system for the positive/negative aspects mentioned in the course end survey and the chapter feedback.

Over a third of the participants explicitly reported to enjoy the interactive exercises and many participants also liked the videos, the programming tasks, and the overall course structure and alignment.

For the negative aspects, many learners reported that they had difficulties solving the programming tasks. Also the rise in difficulty was seen critically and many participants would have liked more comprehensive explanations of the concepts.

There are less categories for the positive aspects compared to the negative aspects. However, since we perform a *qualitative* analysis here, the comparison of numbers is not necessarily meaningful: For the positive aspects, the participants, e.g., usually just reported to like the videos, while for the negative aspects, they explained which aspect about the videos they did not like (which results in several categories). A quantitative comparison of those numbers can sometimes give a hint, but has to be interpreted carefully.

Table 4.2 Categorization of negative aspects mentioned by the participants in the course end survey

2017 (%) (102)	2018 (%) (191)	Negative aspect	Details
0.0	2.6	Automatic feedback	The feedback is not helpful
2.0	0.5	Connections unclear	Connections between the contents elements are not clear
2.9	1.6	Content missing	Some specific content is missing
3.9	2.1	Deadlines	Time-constraints of the course are too strict
1.0	0.0	Discussion board is confusing	Keeping an overview and finding relevant posts is difficult
2.9	2.1	Effort too high	Compared to learning outcome
4.9	4.7	Explanation unclear	
5.9	1.6	External tools	Complicated to use, technical problem with a tool
1.0	0.0	Language problems	German not as mother tongue
19.6	14.1	Level of difficulty increase	
0.0	1.6	Level of difficulty too high	
2.9	0.0	Level of difficulty too low	
2.0	0.5	Miscellaneous	
2.0	0.5	Motivational problem	
2.9	5.8	Obligation to post in forum	Prefer to learn alone without interaction
1.0	1.0	Overview handout is missing	Fact sheet, language reference, etc.
22.5	20.4	Programming too difficult	E.g., difficulties debugging programming errors
1.0	3.7	Quizzes	Questions or answers unclear, too easy
2.0	1.6	Sample solution is missing	Missing or available too late
0.0	1.0	Specific task too difficult	Any specific task (except a programming task)
1.0	1.6	Syntax problems	
4.9	5.2	Task description unclear	
1.0	0.5	Teacher feedback is missing	For drawing or programming exercises
0.0	6.8	Technical problems	Regarding the learning platform, the videos, etc.
11.8	7.9	Too concise	Prefer a more extensive explanation
7.8	4.7	Too few examples	
2.0	2.6	Too few exercises	
1.0	2.6	Too few hints	

(continued)

Table 4.2 (continued)

2017 (%) (102)	2018 (%) (191)	Negative aspect	Details
0.0	0.0	Too few programming tasks	
3.9	5.2	Too theoretical	Prefer more practical explanations and tasks
7.8	7.9	Video style	Too serious, too little enthusiasm, etc.
1.0	0.5	Videos too fast	
1.0	1.6	Videos too long	
0.0	0.5	Videos too short	
1.0	0.0	Videos too slow	
0.0	0.0	Videos volume too low	
0.0	0.0	Weekly exam is missing	Prefer to have additional exams in each chapter

Results for the Chapter Feedback

In the two investigated course runs, we received $255 + 460 = 715$ responses for the chapter feedback as a total their five chapters. The statements are more specific regarding particular aspects of the course sections than in the course end survey. We omit the detailed results for the positive aspects since they mainly reflect the results from the course end feedback and are not very interesting for adapting the course. The results of the qualitative content analysis for the negative aspects are shown in Table 4.3.

Many participants reported that the effort compared to the learning outcome is too low in Chap. 1; they refer to two specific exercises where the learners are supposed to draw a graphic (the layout of their room and an object diagram). This is closely connected to the negative mentioning of embedded tools in Chap. 1.

In Chap. 2, the participants found the content to be too short and too theoretical in the course run 2017, but not in 2018 where we had added another practical exercise. The learners reported problems with an embedded tool in 2018, which was fixed by the authors of the external tool during the course run.

The first programming exercises start in Chap. 3. However, the problems regarding programing tasks are mainly reported in Chap. 4. In this chapter, the tasks are more complex and obviously a hurdle for learners. This coincides with statements regarding a high difficulty, mainly also in Chap. 4.

Table 4.3 lists several more aspects mentioned by the learners, some of which were easy to change in the consecutive course runs, but also some that directly contradict each other.

Results for the Drop-Out Survey

For the drop-out survey, we sent out $1474 + 2270 = 3744$ individualized e-mails to the participants that registered but did not successfully complete the course (course

Table 4.3 Categorization of negative aspects mentioned by the participants in the chapter feedback

2017					2018					Course run
1	**2**	**3**	**4**	**5**	**1**	**2**	**3**	**4**	**5**	**Chapter**
108	**69**	**39**	**28**	**11**	**171**	**147**	**77**	**44**	**21**	**Responses**
0	0	0	0	0	0	0	0	0	0	Automatic feedback
1	2	0	0	0	0	0	0	0	0	Connections unclear
0	1	0	0	0	1	0	0	1	1	Content missing
3	0	0	0	0	0	0	0	1	0	Deadlines
2	0	0	0	0	1	0	0	0	0	Discussion board is confusing
12	0	0	0	0	24	2	0	2	1	Effort too high
1	0	2	2	0	3	2	3	4	1	Explanation unclear
17	0	0	0	0	13	23	1	0	1	External tools
0	0	0	0	0	2	0	0	0	0	Language problems
0	0	0	0	0	0	0	0	0	0	Level of difficulty increase
0	0	1	4	1	2	1	1	11	3	Level of difficulty too high
2	0	0	0	0	4	1	0	0	0	Level of difficulty too low
1	0	0	2	0	3	1	1	0	0	Miscellaneous
0	0	0	0	0	2	0	0	0	0	Motivational problem
0	0	0	0	0	1	0	0	0	2	Obligation to post in forum
2	0	0	0	0	3	0	2	0	1	Overview handout is missing
0	0	1	16	8	0	0	2	20	6	Programming too difficult
1	3	3	0	0	6	11	0	0	0	Quizzes
1	3	0	0	0	1	0	0	0	0	Sample solution is missing
0	1	0	0	0	2	12	0	0	0	Specific task too difficult
0	0	0	0	0	1	0	1	1	0	Syntax problems
7	0	9	1	0	5	0	4	0	0	Task description unclear
6	0	0	0	1	0	0	0	0	0	Teacher feedback is missing
6	0	1	1	0	9	8	12	0	0	Technical problems
2	6	0	0	0	5	3	2	1	0	Too concise
0	2	0	0	0	3	0	0	0	0	Too few examples
1	3	1	0	0	1	0	4	0	0	Too few exercises
0	2	0	0	0	0	2	1	0	0	Too few hints
3	1	0	0	1	0	14	1	0	0	Too few programming tasks
0	5	0	0	0	3	1	0	0	1	Too theoretical
0	1	0	0	0	16	1	0	1	0	Video style
0	1	0	0	0	2	0	0	0	0	Videos too fast
0	0	2	2	0	3	1	0	3	0	Videos too long
1	2	0	0	0	0	0	0	0	0	Videos too short

(continued)

Table 4.3 (continued)

2017					2018					Course run
0	0	0	2	0	2	0	0	0	0	Videos too slow
0	0	0	0	0	1	0	0	0	0	Videos volume too low
3	1	0	0	0	1	0	0	0	0	Weekly exam is missing

runs 2017 and 2018). We received $227 + 411 = 638$ answers regarding the reason for that. The qualitative content analysis yielded 22 categories and the frequencies of the categories are displayed in Table 4.4.

Again we want to note that not everybody not finishing a course is necessarily an actual drop-out since there are several other possible reasons. By far the most prevalent reason mentioned was a lack of time (either due to side jobs, travel or other obligations). Several other participants apparently realized that they were already

Table 4.4 Categorization of reasons to not finish the course mentioned by the participants in the drop-out survey

2017 227	2018 411	Reason for not finishing	Details
1	6	Effort too high	Especially regarding the learning outcome
3	1	Explanations unclear	
5	22	No further need for participation	E.g., due to changes of the study major
12	16	No internet connection	E.g., due to moving or travel
8	10	Illness	
9	14	Motivational problems	
4	13	Level of difficulty too high	
2	0	Level of difficulty too low	
16	26	Level of difficulty increase	
1	3	Only have a look	From the start no intention to learn actively
5	8	Private reasons	
4	12	Programming too difficult	
24	32	Contents already known	
0	6	Miscellaneous	
6	5	Registration too late	
6	4	Language problems	
3	11	Technical problems	E.g., with the learning platform
0	5	Forgotten	Registered but did not think of it afterwards
1	4	Video style	
141	223	Time constraints	Due to jobs, university courses etc.
1	3	Too concise	Explanations too concise to understand
0	4	Too theoretical	Especially regarding the learning outcome

familiar with the contents and, therefore, discontinued the course. Further, both personal (illness, late registration, private reasons, etc.) as well as course-related reasons (rise in difficult, language problems, explanations, etc.) were mentioned. Technical problems hardly appear to be a reason to discontinue with the course.

4.4.5 Adaptation of the Course

Following the design-based research approach, we used the results to adapt the teaching, i.e., our online course. For the adaptation, we considered especially the negative aspects of the course end survey and chapter feedback, but also the positive aspects and the reasons for not finishing the course. The categorization helps to keep an overview, to identify and group similar issues, and also to notice contradicting views. The frequencies of the statements in the categories can also give hints which aspects are urgent for many learners and which statements are only isolated opinions of very few.

Some of the problems and suggestions mentioned in the learners' statements can be addressed easily while others would require big changes to the course structure. We applied some of the changes already during the course runs (i.e., in the case of technical problems or misleading task descriptions) while we changed bigger issues only after the course runs (i.e., introduce completely new exercises). In the following we describe the adaptations we made to the course.

In general, we clarified textual explanations, task descriptions and quizzes wherever reported. We did, however, not change any video recordings yet due to the costs involved.

Since the learners reported problems with an external drawing tool in Chap. 1 of run 2017, we provided alternatives in 2018 and allowed simply draw using paper and upload a photo of the result. We additionally clarified some task descriptions.

Because the learners found Chap. 2 too short and too theoretical, we added another practical exercise (A2.2.6) in 2017 and the feedback in 2018 shows that this is not seen as problem anymore. However, it turned out that the new exercises was seen as difficult, why we added some hints at critical steps after the course run 2018.

In Chap. 4, many learners had difficulties with the actual programming. We therefore added a smaller initial programming task (A4.1.5), provided more explanations and modified the existing programming tasks. Additionally, we included a page giving hints at how to best learn in a self-learning environment like a MOOC and pointing to supporting offers like the discussion board. This had some positive results, but still many learners struggle when having to program a non-trivial class with attributes and methods. After course run 2018 we therefore introduced an explanation of debugging basics, an overview of variable and array instantiation in Java, as well as a series of 20 step-wise hints for one programming exercise.

Most learners who did not finish the course mentioned time-constraints as the main reason. We therefore also changed the course organization slightly: In 2017, the chapters were released on Monday and due the following Monday. In 2018, the chapters were already released on the Friday before. This made it possible that each chapter is available on two weekends with a small overlap of the chapters.

We did, of course, not apply all changes proposed by the participants. So far we focused on changes that are not controversial among the learners and do not require to change the overall course structure.

4.4.6   Learners' Communication

Unlike in the pilot course run 2016 (Krugel and Hubwieser 2017), there were lively discussions in the discussion boards of course runs 2017 and 2018. At the end of the five-week courses, the discussion boards of course runs 2017 and 2018 contained $72 + 122 = 194$ topics with $197 + 244 = 441$ replies. The exercise-specific forums contained $1068 + 2442 = 3510$ topics with $288 + 932 = 1220$ replies in total. This is presumably partly due to our intensified aims at encouraging asking questions and helping other students. Another reason is certainly the substantially higher number of participants because discussions take place much easier above a certain critical mass.

During the course we observed the discussion board daily and reacted accordingly when needed. However, we did not systematically analyze the contents of the discussions yet.

4.4.7   Workload

In the course end survey, the learners reported their average weekly workload. The average was 3.8 h per week in 2017 and 4.6 h per week in 2018. This increase is presumably due to the additional exercises and material. Another reason might be that more learners reached the later and reportedly more time consuming chapters of the courses. Some participant's reported in the discussion forum and feedback texts to spend more than 15 h per week on the course, especially when stuck in programming exercises.

4.4.8   Learners' Performance

In principle, MOOCS are intended to attract large numbers of participants. In most cases, they don't offer any tutoring by human individuals. In consequence, there is no way of monitoring the atmosphere and the learning behavior of the participants by personal contact with teaching personal. Thus, it is seems essential to monitor the behavior of the learners by automatic means to get necessary feedback on particular difficult content and drop-out rates or reasons. Due to the large scale of the courses, the behavior of some few individuals is not very relevant. Instead, we are interested in the learning process of larger groups that share certain behavior features, e.g., regarding scoring of certain tasks or drop-out points, which seems to be a classical application for statistical cluster analysis.

First, we performed a cluster analysis of the pilot conduction of LOOP in 2016, for which we calculated the average of the achieved relative scores over each of the 16 course sections for each of the 187 participants. The results reflected in a quite instructive way, how the performance of large groups of participants developed over these sections and in which sections a substantial number of them gave up. For the next course runs in 2017 and 2018, the number of participants increased substantially, which allows us to perform cluster analysis on the level of singles tasks to get a much more detailed picture of the performance development.

For this purpose, we conducted a hierarchical cluster analysis on the individual scores of all tasks of the course. In a first step, we cleaned raw score matrices (one line per individual participant and one column by each task) by removing all lines that contained exclusively empty values ("not attempted"). In a second step, we normalized the resulting matrices to an interval scale (from 0.00 to 1.00) by replacing all "not attempted" values by 0 and by dividing all score values by the maximum score of the respective tasks (columns).

For the clustering, we regarded the columns of this matrix as dimensions. Thus, the set of all scores of each participant could be interpreted as the definition of a certain position in a multidimensional space. By this way, the pairwise distance between the positions of all participants could be calculated in a specific distance matrix. Looking for the best result, we tried two different distance metrics (Maximum and Euclidian). Finally, a hierarchical clustering was performed on this distance matrix, starting with one cluster per person and combining successively more and more persons to larger clusters according to their relative distance, applying several different clustering strategies [ward.D, Complete, Average and McQuitty, for details see (Everitt et al. 2001)]. The calculation was performed in the statistical programming language R by applying the function *hclust*.

As hierarchical clustering is a local heuristic strategy, the results have to be inspected according their plausibility. For this purpose, we looked for plausible dendrograms that represented a proper hierarchy. We found that the Euclidean distance metrics produced the best results in combination with the ward.D algorithm. Figure 4.2 shows an exemplary dendrogram for this combination. To find a suitable number of clusters, we inspected these dendrograms from the top down to a level where we found as many clusters as possible, but avoiding too small clusters with too small number of members. We found that the best height to cut would be at 4 branches, which suggests the following clusters c1, c2, c3, and c4 in course run 2017 and d1, d2, d3, and d4 in course run 2018 (with the numbers of members in parentheses):

- Course run 2017: c1 (104), c2 (207), c3 (235), and c4 (328)
- Course run 2018: d1 (513), d2 (196), d3 (472), and d4 (469).

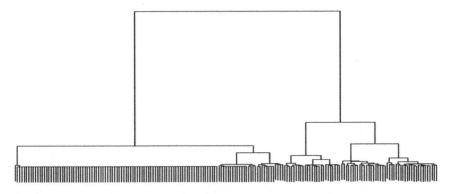

Fig. 4.2 Exemplary dendrogram, clustering by *Euclidean* distance with the *ward.D* algorithm

Finally, we calculated the average performance over all course tasks for each of the 4 clusters. The results are displayed in Fig. 4.3 (for course run 2017) and Fig. 4.4 (for course run 2018). Please note that there is a small difference in the task lists between the runs of 2017 and 2018, because in 2018 two new exercises were added (A2.2.5 and A4.1.5). Exercise A2.2.5 (Maze) has been changed for a more flexible answer format and is now obviously more difficult. In addition, some other exercises might be easier due to improved explanations, etc.

Due to its design, our clustering reflects the performance and drop-out behavior of four typical groups that represent many participants each. Unfortunately, the numbering of the clusters is set arbitrarily by the R packages. To support the comparison between the two runs, we introduce new names for the clusters:

1. The "high performers" kept comparably high scores over the whole course in both runs (c1 and d2).

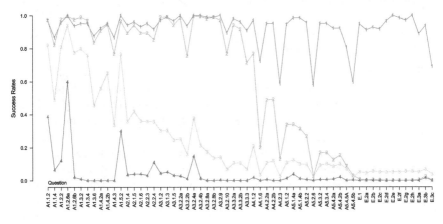

Fig. 4.3 Average performance of the learner clusters c1, c2, c3, and c4 in the course sections in 2017

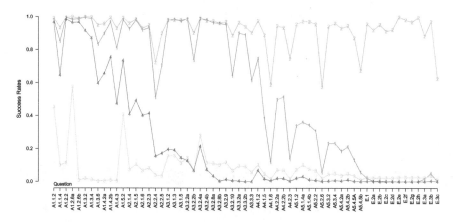

Fig. 4.4 Average performance of the learner clusters d1, d2, d3, and d4 in the course sections in 2018

Table 4.5 Clusters of participants in the course runs 2017 and 2018

	2017		2018	
Name	Cluster	Members (%)	Cluster	Members (%)
High performers	c1	11.9	d2	11.9
Late droppers	c2	23.7	d1	31.1
Early droppers	c3	26.9	d4	28.4
Low performers	c4	37.5	d3	28.6

2. The "late droppers" started with high scores, but dropped starting from Sect. 4.4 (c2 and d1).
3. The performance of the "early droppers" dropped already during Sect. 4.1 (c3 and d4).
4. The "low performers" reached high scores only in a few tasks (1.6.2a, 1.5.2, and 3.2.4a), but performed low in the rest (c4 and d3).

The clusters are summarized in Table 4.5. A comparison of the numbers of members of these clusters demonstrates that the percentage of high performers remained constant. The number of late droppers increased significantly, while the low performers were reduced. This might be interpreted as an overall improvement of LOOP.

In the course run 2017, 113 of the 2390 (4.7%.) registered participants successfully passed the course. In 2018, this ratio was up to 213 out of 2520 (8.5%). This is quite an encouraging result and indicates that the adaptations of the course meet the learners' preferences and help with the challenges.

4.5 Discussion

Using a mixed-methods approach and following a design-based research method-ology, we gained insights into the perspectives of learners in MOOCs, which also served as basis for adapting the course.

4.5.1 Lessons Learned

1. What are learners' preferences in self-regulated introductory programming MOOCs?

Learners reported to like the very clear structure of a MOOC with well-aligned videos, quizzes, and exercises. Especially many learners mention to enjoy interactive exercises that let them interact with the concepts directly.

2. What are the main challenges learners face in self-regulated introductory programming MOOCs?

A major challenge for learners is to handle a high level of difficulty, especially when the difficulty increases during the course. This is also closely connected to motivational factors of the learners. If it is not possible to avoid such an increase, it seems to help the participants if the rise in difficulty is explicitly announced before and several offers for assistance are clearly pointed out.

Interestingly, some participants explicitly stated to dislike engaging in discussion boards and prefer to study on their own. This should be kept in mind when designing a course with collaborative elements, compulsory peer-feedback etc.

It was also confirmed to be important to keep the technical barriers very low when offering a course which is available worldwide for everybody. Especially novices cannot be expected to install additional software like an integrated development environment (IDE) or compilers when sitting alone at their computer. We, therefore, integrated only purely web-based tools and a few participants still mentioned some technical difficulties (even though no severe problems). In the recent course runs, several participants mentioned they would like to complete the course on their mobile phone or tablet computer. Even though the learning platform already provides apps for the most popular operating systems, some exercises involving external tools to not work satisfactorily.

3. What are the main reasons for not finishing introductory programming MOOCs?

Time-wise flexibility seems to be very important to participants of online courses. For each chapter in the course run 2018, we allocated 10 days including two weekends in which the participants could work on the content at their own pace. However, the vast majority of participants that did not complete the course, still mentioned time constraints as the main reason. This confirms the main findings by Eriksson et al. (2017) and Topali et al. (2019).

Our qualitative analysis resulted in 22 categories of reasons for not finishing a MOOC. The interview study of Zheng et al. (2015) yielded 8 categories and the interview study of Eriksson et al. (2017) found 11 main factors (grouped into 4 themes). Most of our categories can be mapped to one of the categories of the previous studies and vice-versa. Table 4.6 shows a comparison of the three category systems. Our categorization is a bit more fine-granular and also more specific for programming MOOCs.

4.5.2 Limitations

The learners' feedback of one MOOC is the main basis of our analysis. By its very nature, feedback is only self-reported and, therefore, not necessarily objective. For example, when asking for the reasons to not finish the course, participants might attribute their drop-out rather to time constraints than to their inability to solve the tasks. Therefore, it seems very important to follow a mixed-methods approach also in the future.

In the course runs, we did not study the effect of single changes to the course in isolation. This is organizationally not very easy but we also consider running controlled AB-tests in the future.

4.5.3 Outlook

The enhanced completion rate during the three course runs is presumably mostly due to our modifications on the course based on the learners' feedback. This underlines the necessity of a participatory approach to the ongoing development of online courses and teaching in general.

In the future, we plan to analyze further data of our rich data collection. We are currently analyzing students' solutions of programming exercises of an on-campus university course as basis for the definition of competencies for OOP and the automatic generation of feedback (Krugel et al. 2020); we plan to extend this analysis to the programming solutions of our MOOC as well. Furthermore, we will take a closer look at the discussion board questions which can give insights into, e.g., misconceptions of the learners.

We have just added another chapter to the course: Chap. 6 introduces inheritance and polymorphism and thereby rounds off the course to cover the most important object-oriented concepts. Additionally, we develop and adapt the course further based on the learners' feedback and the quantitative data.

In the long term, we aim to use LOOP as a general tool to analyze learning processes in object-oriented programming. The online setting allows to perform

Table 4.6 Comparison of reasons for not finishing a MOOC

This work	Zheng et al. (2015)	Eriksson et al. (2017)
Effort too high	High workload	Utilitarian motivation
Explanations unclear		The learner's perception of the course design
No further need for participation	Learning on demand	Utilitarian motivation
No internet connection		Internet access
Illness		External factors
Motivational problems	Lack of pressure, no sense of community or awareness of others, social influence	Utilitarian motivation, Enjoyment motivation, Study techniques
Level of difficulty too high	Challenging course content	Perceived level of difficulty of the content
Level of difficulty too low		
Level of difficulty increase	Challenging course content	Perceived level of difficulty of the content
Only have a look	Learning on demand	Utilitarian motivation
Private reasons		External factors
Programming too difficult	Challenging course content	Perceived level of difficulty of the content
Contents already known	Learning on demand	Mismatch between expectations and actual content
Miscellaneous		
Registration too late	Lengthy course start-up	
Language problems		English proficiency
Technical problems		
Forgotten	Lengthy course start-up	
Video style		The learner's perception of the course design
Time constraints	Lack of time	Lack of time
Too concise	Challenging course content	The learner's perception of the course design
Too theoretical	Challenging course content	The learner's perception of the course design

experiments and analyses on a scale much larger than in a regular classroom course and, furthermore, poses new research questions (Settle et al. 2014).

Acknowledgements This work was partially supported by the Dr.-Ing. Leonhard Lorenz-Stiftung (Grant No. 949/17) and the European Union's Horizon 2020 research and innovation programme (grant agreement No. 787476). We thank Alexandra Funke and Marc Berges for developing parts of the course. We would furthermore like to thank all course participants for taking the time to provide their valuable feedback and the reviewers for the comments that helped to improve the exposition.

References

Alario-Hoyos, C., Delgado Kloos, C., Estévez-Ayres, I., Fernández-Panadero, C., Blasco, J., Pastrana, S., et al. (2016). Interactive activities: The key to learning programming with MOOCs. In *European Stakeholder Summit on experiences and best practices in and around MOOCs (EMOOCS'16)*. Books on Demand.

Alonso-Ramos, M., Martin, S., Albert Maria, J., Morinigo, B., Rodriguez, M., Castro, M., & Assante, D. (2016). Computer science MOOCs: A methodology for the recording of videos. In *IEEE Global Engineering Education Conference (EDUCON'16)*.

Anderson, T., & Shattuck, J. (2012). Design-based research. *Educational Researcher, 41*(1), 16–25. https://doi.org/10.3102/0013189X11428813.

Bajwa, A., Hemberg, E., Bell, A., & O'Reilly, U.-M. (2019). Student code trajectories in an introductory programming MOOC. In Unknown (Ed.), *Proceedings of the Sixth (2019) ACM Conference on Learning @ Scale—L@S '19* (pp. 1–4). ACM Press. https://doi.org/10.1145/3330430.333 3646.

Cooper, S., & Cunningham, S. (2010). Teaching computer science in context. *ACM Inroads, 1,* 5–8. https://doi.org/10.1145/1721933.1721934.

Crues, R. W., Bosch, N., Anderson, C. J., Perry, M., Bhat, S., & Shaik, N. (2018). Who they are and what they want: Understanding the reasons for MOOC enrollment. In *Proceedings of the 11th International Conference on Educational Data Mining, EDM 2018, Buffalo, NY, USA, July 15–18, 2018*. https://educationaldatamining.org/files/conferences/EDM2018/papers/EDM2018_paper_121.pdf.

Delgado Kloos, C., Munoz-Merino, P. J., Munoz-Organero, M., Alario-Hoyos, C., Perez-Sanagustin, M., Parada G., H. A., et al. (2014). Experiences of running MOOCs and SPOCs at UC3M. In *IEEE Global Engineering Education Conference (EDUCON'14)*.

Derval, G., Gego, A., Reinbold, P., Frantzen, B., & van Roy, P. (2015). Automatic grading of programming exercises in a MOOC using the INGInious platform. In *European Stakeholder Summit on experiences and best practices in and around MOOCs (EMOOCS'15)*.

Design-Based Research Collective. (2003). Design-based research: An emerging paradigm for educational inquiry. *Educational Researcher, 32*(1), 5–8. https://doi.org/10.3102/0013189X0320 01005.

Eriksson, T., Adawi, T., & Stöhr, C. (2017). "Time is the bottleneck": A qualitative study exploring why learners drop out of MOOCs. *Journal of Computing in Higher Education, 29*(1), 133–146. https://doi.org/10.1007/s12528-016-9127-8.

Estler, C., & Nordio, M., *Codeboard*. https://codeboard.io/.

Everitt, B. S., Landau, S., & Leese, M. (2001). *Cluster analysis*. Arnold.

Falkner, K., Falkner, N., Szabo, C., & Vivian, R. (2016). Applying validated pedagogy to MOOCs. In *ACM Conference on Innovation and Technology in Computer Science Education (ITiCSE'16)* (pp. 326–331). ACM. https://doi.org/10.1145/2899415.2899429.

Fitzpatrick, J. M., Lédeczi, Á., Narasimham, G., Lafferty, L., Labrie, R., Mielke, P. T., et al. (2017). Lessons learned in the design and delivery of an introductory programming MOOC. In M. E. Caspersen, S. H. Edwards, T. Barnes, & D. D. Garcia (Eds.), *Proceedings of the 2017 ACM SIGCSE Technical Symposium on Computer Science Education—SIGCSE '17* (pp. 219–224). ACM Press. https://doi.org/10.1145/3017680.3017730.

Garcia, F., Diaz, G., Tawfik, M., Martin, S., Sancristobal, E., & Castro, M. (2014). A practice-based MOOC for learning electronics. In *IEEE Global Engineering Education Conference (EDUCON'14)*.

Geldreich, K., Simon, A., & Hubwieser, P. (2019). Design-Based Research als Ansatz zur Einführung von Algorithmik und Programmierung an bayerischen Grundschulen. *MedienPädagogik: Zeitschrift Für Theorie Und Praxis Der Medienbildung, 33* (Medienpäda). https://doi.org/10.21240/mpaed/33/2019.02.15.X.

Gries, D. (2008). A principled approach to teaching OO first. *ACM SIGCSE Bulletin, 40*(1), 31. https://doi.org/10.1145/1352322.1352149.

Guo, P. J., Kim, J., & Rubin, R. (2014). How video production affects student engagement. In M. Sahami, A. Fox, M. A. Hearst, & M. T. H. Chi (Eds.), *1st ACM Conference on Learning@Scale (L@S'14)* (pp. 41–50). ACM. https://doi.org/10.1145/2556325.2566239.

Hicken, A. (2018). *2019 eLearning hype curve predictions*. https://webcourseworks.com/elearning-predictions-hype-curve/.

Hubwieser, P. (2008). Analysis of learning objectives in object oriented programming. In R. T. Mittermeir & M. M. Syslo (Eds.), *Lecture notes in computer science, Informatics Education—Supporting Computational Thinking, 3rd International Conference on Informatics in Secondary Schools—Evolution and Perspectives (ISSEP'08)* (pp. 142–150). Springer. https://doi.org/10.1007/978-3-540-69924-8_13.

Hubwieser, P., Giannakos, M. N., Berges, M., Brinda, T., Diethelm, I., Magenheim, J., et al. (2015). A global snapshot of computer science education in K-12 schools. In *ITiCSE Working Group Reports* (pp. 65–83). ACM. https://doi.org/10.1145/2858796.2858799.

Krugel, J., & Hubwieser, P. (2017). Computational thinking as springboard for learning object-oriented programming in an interactive MOOC. In *2017 IEEE Global Engineering Education Conference (EDUCON)* (pp. 1709–1712). IEEE. https://doi.org/10.1109/EDUCON.2017.7943079.

Krugel, J., & Hubwieser, P. (2018). Strictly objects first: A multipurpose course on computational thinking. In M. S. Khine (Ed.), *Computational thinking in the STEM disciplines* (Vol. 49, pp. 73–98). Springer International Publishing. https://doi.org/10.1007/978-3-319-93566-9_5.

Krugel, J., Hubwieser, P., Goedicke, M., Striewe, M., Talbot, M., Olbricht, C., et al. (2020). Automated measurement of competencies and generation of feedback in object-oriented programming courses (preprint). In *2020 IEEE Global Engineering Education Conference (EDUCON)*. IEEE.

Kurhila, J., & Vihavainen, A. (2015). A purposeful MOOC to alleviate insufficient CS education in Finnish schools. *ACM Transactions on Computing Education, 15*(2), 1–18. https://doi.org/10.1145/2716314.

Liyanagunawardena, T. R., Lundqvist, K. O., Micallef, L., & Williams, S. A. (2014). Teaching programming to beginners in a massive open online course. In *Building Communities of Open Practice (OER'14)*.

Luik, P., Feklistova, L., Lepp, M., Tõnisson, E., Suviste, R., Gaiduk, M., et al. (2019). Participants and completers in programming MOOCs. *Education and Information Technologies, 24*(6), 3689–3706. https://doi.org/10.1007/s10639-019-09954-8.

Luik, P., Suviste, R., Lepp, M., Palts, T., Tõnisson, E., Säde, M., & Papli, K. (2019). What motivates enrolment in programming MOOCs? *British Journal of Educational Technology, 50*(1), 153–165. https://doi.org/10.1111/bjet.12600.

Mayring, P. (2000). Qualitative content analysis. *Forum Qualitative Sozialforschung/Forum: Qualitative Social Research, 1*(2), Article 20. https://nbn-resolving.de/urn:nbn:de:0114-fqs0002204.

Moreno-Marcos, P. M., Muñoz-Merino, P. J., Maldonado-Mahauad, J., Pérez-Sanagustín, M., Alario-Hoyos, C., & Delgado Kloos, C. (2020). Temporal analysis for dropout prediction using self-regulated learning strategies in self-paced MOOCs. *Computers & Education, 145*, 103728. https://doi.org/10.1016/j.compedu.2019.103728.

Papavlasopoulou, S., Giannakos, M. N., & Jaccheri, L. (2019). Exploring children's learning experience in constructionism-based coding activities through design-based research. *Computers in Human Behavior, 99*, 415–427. https://doi.org/10.1016/j.chb.2019.01.008.

Piccioni, M., Estler, C., & Meyer, B. (2014). SPOC-supported introduction to programming. In Å. Cajander, M. Daniels, T. Clear, & A. Pears (Chairs), *Innovation & Technology in Computer Science Education (ITiCSE'14)*, Uppsala, Sweden.

Plomp, T. (2007). Educational design research: An introduction. *An Introduction to Educational Design Research*, 9–35.

Reeves, T. C. (2006). Design research from a technology perspective. In J. van den Akker (Ed.), *Educational design research* (pp. 52–66). Routledge.

Settle, A., Vihavainen, A., & Miller, C. S. (2014). Research directions for teaching programming online. In *Proceedings of the International Conference on Frontiers in Education Computer Science and Computer Engineering (FECS'14)*.

Skoric, I., Pein, B., & Orehovacki, T. (2016). Selecting the most appropriate web IDE for learning programming using AHP. In *39th International Convention on Information and Communication Technology, Electronics and Microelectronics (MIPRO'16)* (pp. 877–882). IEEE. https://doi.org/10.1109/MIPRO.2016.7522263.

Staubitz, T., Klement, H., Teusner, R., Renz, J., & Meinel, C. (2016). CodeOcean—A versatile platform for practical programming excercises in online environments. In *IEEE Global Engineering Education Conference (EDUCON'16)*.

Striewe, M., & Goedicke, M. (2013). JACK revisited: Scaling up in multiple dimensions. In *Lecture notes in computer science, 8th European Conference, on Technology Enhanced Learning (EC-TEL'13): Scaling up Learning for Sustained Impact* (pp. 635–636). Berlin, Heidelberg: Springer. https://doi.org/10.1007/978-3-642-40814-4_88.

Topali, P., Ortega-Arranz, A., Er, E., Martínez-Monés, A., Villagrá-Sobrino, S. L., & Dimitriadis, Y. (2019). Exploring the problems experienced by learners in a MOOC implementing active learning pedagogies. In M. Calise, C. Delgado Kloos, J. Reich, J. A. Ruiperez-Valiente, & M. Wirsing (Eds.), *Lecture notes in computer science. Digital Education: At the MOOC Crossroads Where the Interests of Academia and Business Converge* (Vol. 11475, pp. 81–90). Springer International Publishing. https://doi.org/10.1007/978-3-030-19875-6_10.

Vihavainen, A., Luukkainen, M., & Kurhila, J. (2012). Multi-faceted support for MOOC in programming. In R. Connolly (Ed.), *ACM Digital Library, Proceedings of the 13th Annual Conference on Information Technology Education* (p. 171). ACM. https://doi.org/10.1145/2380552.2380603.

Wing, J. (2006). Computational thinking. *Communications of the ACM, 49*(3), 33–35. https://doi.org/10.1145/1118178.1118215.

Zheng, S., Rosson, M. B., Shih, P. C., & Carroll, J. M. (2015). Understanding student motivation, behaviors and perceptions in MOOCs. In D. Cosley, A. Forte, L. Ciolfi, & D. McDonald (Eds.), *Proceedings of the 18th ACM Conference on Computer Supported Cooperative Work & Social Computing—CSCW '15* (pp. 1882–1895). ACM Press. https://doi.org/10.1145/2675133.2675217.

Johannes Krugel studied computer science with minor Psychology at the Free University of Berlin. Since 2009–2016, he was teaching and research assistant at the Technical University of Munich. In 2016, he received his Dr. rer. nat. in Theoretical Computer Science at the Technical University of Munich with his dissertation on efficient algorithms and data structures for strings. For several years he was responsible for the education, support, and supervision of the teaching assistants at the Faculty for Computer Science. Since 2016, he is postdoctoral researcher in the working group for Didactics in Computer Science of Prof. Hubwieser. His research focus is on computer science education and computer-based learning, especially in higher education. He designed the MOOC and coordinated the development and implementation.

Peter Hubwieser was teaching Mathematics, Physics, and Computer Science at Bavarian Gymnasiums for 15 years. In 1995, he received his Dr. rer. nat. in Theoretical Physics at the Ludwig Maximilian University of Munich. From 1994 to 2002, he has been delegated to the Faculty of

Informatics of the Technical University of Munich on behalf of the Bavarian Ministry of Educa-tion, in charge for the implementation of new courses of studies for teacher education and a compulsory subject of Computer Science at Bavarian secondary schools. Since June 2002, he is working as an associate professor at the Technical University of Munich. From 2002 to 2015, he was additionally visiting Professor at the Alpen-Adria-University of Klagenfurt, 2007 at the University of Salzburg, and 2008 at the University of Salzburg. In 2006, he received the Bavarian State Award for Education and Culture and 2011 the Second place award in the nation-wide contest for Subject Matter Didactics of the Polytechnische Gesellschaft Frankfurt. Since 2009, he holds the position of the Information Officer of the TUM School of Education. In 2015, he took over the position of the scientific director of the Schülerforschungszentrum Berchtesgaden additionally to his professorship.

Part III
Novel Frameworks and Pedagogical Considerations

This part provides insight into different pedagogical frameworks, affordances and considerations applied in science learning.

Chapter 5
Music and Coding as an Approach to a Broad-Based Computational Literacy

Michael S. Horn, Amartya Banerjee, and Melanie West

Abstract This chapter takes up two questions related to science learning in the twenty-first century. How do we develop broad-based computational literacy skills for the next generation of learners? And, how do we do that in a way that engages diverse learners whose voices have been historically marginalized in computing fields? To think about these questions, we provide a case study of student learning around music and coding in the context of a middle school summer camp. We reflect on the process through which multiple literacies (music as a literacy and computational literacy) shape student learning, creative expression, and engagement. We propose that developing computational literacy skills for the purposes of science might best be accomplished through a long-term, multidisciplinary approach in which students engage in many kinds of activities for diverse purposes, including that of personal creative expression. Music, in turn, provides a particularly rich context through which to explore concepts of computer programming.

5.1 Introduction

This chapter takes up what we see as two critical questions related to science learning in the twenty-first century. First, how do we develop broad-based computational literacy skills for the next generation of learners? And, second, how do we do that in a way that is inclusive—in a way that engages diverse learners including women and students of color whose voices have been historically marginalized in computing fields? The first question is important because the work of scientific inquiry is increasingly computational in nature. Steady advances in processing speeds, storage capacity, connectivity, and algorithmic sophistication have created powerful new tools for understanding phenomena across a wide range of scientific disciplines. The opening decades of this century, in particular, have witnessed a radical reorganization of both the natural and social sciences. Early examples such as the sequencing of the

M. S. Horn (✉) · A. Banerjee · M. West
Northwestern University, Evanston, IL, USA
e-mail: michael-horn@northwestern.edu

© Springer Nature Singapore Pte Ltd. 2020
M. Giannakos (ed.), *Non-Formal and Informal Science Learning
in the ICT Era*, Lecture Notes in Educational Technology,
https://doi.org/10.1007/978-981-15-6747-6_5

human genome (IHGSC 2001) or innovations in computational methods in quantum chemistry (Pople 2003; Kohn 2003) have given way to a sea change in innovation with computing that has supported more recent breakthroughs such as the detection of gravitational waves (Abbot et al. 2016) or the first image of a black hole captured from telescope data (Event Horizon Telescope Collaboration 2019).

The second question is important because our education systems continue to struggle to engage underrepresented groups in computational fields, particularly along racial, ethnic, and gender lines. Recent data from high school advanced placement exams in the United States gives us some reason to hope that the situation is slowly starting to improve,[1] but surveys of college-level computing degree programs also make it painfully clear that we have much more work to do to ensure a diverse computational future. As just one example, according to data from the Computing Research Association's Taulbee survey women made up 18.3% of Ph.D.'s awarded in Computer Science in the United States and Canada in 2016, while underrepresented minority students accounted for slightly less than 2% (Zweben and Bizot 2018).

In this chapter, we provide a case study of student learning around music and computer programming. We consider interviews, observations, and artifacts created by middle school students in a summer camp setting using a free, online platform called TunePad that our team has been developing for the past three years as part of a larger research project funded by the US National Science Foundation.[2] We reflect on the process through which multiple literacies (music and code) shape student learning, creative expression, and engagement.

The domain of music might seem like an odd fit for an edited volume on the topic of science learning. Our argument, however, is that if we take seriously the idea that computation is a form of literacy, then we must also understand that the process of becoming literate isn't something that happens overnight—it plays out over a period of many years across many different kinds of learning spaces, both formal and informal. The notion of literacy also implies a set of social, cognitive, and technical skills that are relevant to more than one domain of human activity (diSessa 2018; Vee 2017). In essence, we're arguing that developing computational literacy skills for the purposes of science might best be accomplished through a long-term, multidisciplinary approach in which students engage in many kinds of activities for diverse purposes, including that of personal creative expression. As we hope to demonstrate in this chapter, music provides a particularly rich context through which to explore concepts of computer programming.

---

[1] https://home.cc.gatech.edu/ice-gt/597.

[2] This work is supported by the National Science Foundation grants 1612619, 1451762, and 1837661. Any opinions, findings and conclusions or recommendations expressed in this material are those of the authors and do not necessarily reflect the views of the National Science Foundation.

5.2 Setting the Stage: Computation as a Literacy

We don't normally think of it as such, but writing is a technology, which means that a literate person is someone whose thought processes are technologically mediated. We became cognitive cyborgs as soon as we became fluent readers, and the consequences of that were profound.

—Ted Chiang, The Truth of Fact, the Truth of Feeling (2019)

The perspective of computation as a form of literacy is gaining increasing acceptance, especially through the work of scholars such as diSessa (2018), Vee (2017), and others. Computational literacy can be thought of as a kind of "material intelligence" rooted in representational forms (such as programming languages) as well as the technological means for producing, manipulating, and communicating with those forms. It is dependent on individual cognitive skills as well as communities of practice within which with those skills are valued as a means of expression and thought. This definition of computational literacy (CL) contrasts with definitions of computational thinking (CT) as an approach to problem-solving and design that draws directly on concepts, skills, and tools from the field of Computer Science (Brennan and Resnick 2012; Grover and Pea 2013; Wing 2006). Both CL and CT emphasize the transformational power of computation on human thought. However, the CL perspective gives us a useful vision for what the process of becoming computationally literate might look like: an inherently social endeavor that plays out over a long period of time in a variety of settings with a range of materials. Importantly, becoming literate also involves a transformation of participation within various communities of practice. As Kafai and Burke (2013) describe it, "learning to code has shifted from being a predominantly individualistic and tool-oriented to one that is decidedly sociologically and culturally grounded in the creation and sharing of digital media" (p. 603). Correspondingly, this process also involves shifts in personal identity (that of becoming literate) and can be applied to a diverse range of social needs, including communication, artistic expression, entertainment, and problem-solving.

Before we get into music and our summer camps, it will be helpful to take a detour to understand what computational literacy looks like in a professional context. Below we share a short excerpt from a lengthy interview that one member of our team conducted with an astrophysics researcher studying the evolution of neutron star systems. This interview was collected as part of a larger research project designed to understand the computational, tools, methods, and practices of professional researchers across a range of scientific disciplines. In the interviews, we asked participants to describe their research interests and activities. We then used semi-structured prompts to probe how computational tools shape their work. This study is described in more detail in Weintrop et al. (2016).

In this particular interview, the participant was describing the methods he was using to understand the formation of binary neutron star systems and why such "twin" stars often have roughly equal masses.

So what I do is I generate models from stellar evolution codes. So my-my project uses I think three or four different codes. So and then the solar evolution code will output different snapshots of the star at different times so I can grab a snapshot of the star at the main sequence, red giant up to supernova. So what I do is I take one of those snapshots and then I uh, well I pretty much just double it because we're intr- we're interested in simulating twin stars and then I'll put them in my SPH program and then just let them evolve and see what happens…

The language of computational astrophysics takes some getting used to, but one thing that was clear throughout this interview was the pervasive role of computational modeling and simulation in the research (e.g., "stellar evolution codes" and "SPH [Smoothed Particle Hydrodynamics] program"), to the point that it was extremely difficult to tell when he was talking about reality, as observed telescope data, and when he was talking about the output of computational simulations—or whether such a distinction between realities even makes sense to consider in the context of his research. As the interview progressed, we interrupted to ask if he ever used tools to visualize his data. He responded by saying

> I mean it's just graphs of you know the different compositions. But what I do is o-out of curiosity I just wrote up a code to output you know the-the graphs throughout time. So I mean that's not difficult to do because there's so much output that it's easy to just program up something in Python or IDL just to show something that's visual."
>
> (underlines added for emphasis)

What's striking about his response is the fluency with which he makes use of computational tools. He does things that are "easy" to do just "out of curiosity," while showing adaptability with the use of a variety of computational tools ("in Python or IDL") or presumably any number of other tools that he could pick up on an as-needed basis. He also sees an abundance of data as a clear advantage ("there's so much output that it's easy") rather than an obstacle to overcome. This is a clear example of what we think of as a computationally literate professional whose flexible and confident use of computation methods empowers his research.

5.3   Three Less Obvious Implications of Computation as a Literacy

Our interview with this astrophysicist (and others interviews that we have collected like it), set a high bar for the fluency needed for professional practice—this is what computational literacy *can* look like for people in a range of professions, not just scientific research. If we hope to empower future scientists, setting strong but achievable goals in terms of computational literacy will be important. But it's important to realize that a literacy frame for computing is not without complication. In this section, we briefly consider some of the implications that the term *literacy* brings with it from a historical perspective.

The first implication has to do with diSessa's argument that, by definition, literacy is something that operates at the level of a society with broad-ranging applications across many domains of human activity. Or, as Vee writes, "because programming is so infrastructural to everything we say and do now, leaving it to computer science is like leaving writing to English or other language departments" (Vee 2017, p. 7). diSessa (2018) is more pointed in his critique of a computational literacy as *owned* by the field of Computer Science. He recognizes the power of literacy to disrupt and transform culture, while simultaneously lamenting the techno-centric nature of CS education programs and coding bootcamps. What both diSessa and Vee are pointing out is that, in many ways, current CS education campaigns are just that "*computer science*" education campaigns that ostensibly claim to be preparing learners to become computer scientists or software engineers. Prominent examples include CS for All initiatives rolling out at city, state, and national levels and through funding agencies both public and private. We note here that there's nothing wrong with taking a Computer Science course in school. What diSessa takes issue with is the idea that we equate CS with many learning activities that involve developing computational literacy that fall far outside the academic domain of Computer Science.

From a perspective of diversity and inclusion, this raises a critical and inescapable problem—Computer Science is historically (though not exclusively, of course) a field created by culturally homogeneous group of people. This is not to belittle the accomplishment of CS pioneers, nor is it meant to downplay the roles of the many diverse players in constructing the field. It is, however, an observation that the cultural background of a scientific community will indelibly shape the form that a field takes, including everything from the representation systems invented, the people that the field lionizes, and the subtle jokes, humor, and cultural norms that define what it means to be an insider in that field. An inevitable difficulty of CS education programs that build on this culture or that aspire to launch people into CS careers is that sooner or later they run into CS dominant culture. This doesn't mean that culture can't be changed to become more inclusive, but it's still important to recognize cultural baggage for what it is, even if it's not always easy to see from the inside.

A second implication of a literacy perspective relates to the societal power dynamics that literacy creates. Vee reflects on the emergence of widespread literacy in Medieval Europe and draws parallels to the twenty-first century. She notes that societies become dependent on literacy (e.g., bureaucratic systems based on written records and official documents) long before literacy at an individual level becomes widespread. This creates a power differential; those who are literate wield distorted levels of influence over those who are not. The parallels to the tech industry in the twenty-first century aren't difficult to imagine. Those who have the ability to create technology and develop algorithms have an outsized influence over society as a whole. Vee also notes that adopting the term literacy constructs the inverse concept of *illiteracy*. "Because of literacy's heritage of moral goodness, calling something a literacy raises the stakes for acquiring that knowledge. To lack the knowledge, one can be penalized for the immorality of illiteracy–for dragging society down." (Vee 2017, p. 2).

A third implication follows from the second. At many points in history, literacy and the educational systems that supported it have been used as a tool to maintain social, class, and economic power structures along with systems of inequity that reinforce those structures (Vee 2017). Put bluntly, literacy can be as much a tool of oppression as it is empowerment depending on how it is imposed or denied. There are many examples from colonial societies in which those who wield power have dictated what counts as literacy while simultaneously repressing or marginalizing indigenous languages and cultural knowledge systems. Those societies have also denied literacy to segments of populations while using literacy (or illiteracy) as a tool of disenfranchisement. Understanding that literacy has historically been both a tool of oppression and empowerment, we can see it as a wedge that reinforces and widens power structures. It's worth noting that the dominant rhetorical scripts in the Computer Science education community emphasize the need for equity and inclusion (echoing historical movements such as progressive education reform of the nineteenth and twentieth century). But, there is also a strong degree of cognitive dissonance when broader economic, education, and legal systems actively repress and disenfranchise minoritized learners, particularly in countries like the United States (e.g., Margolis 2010). We'll return to these themes again at the end of the chapter to reflect on ways to build computational literacy for diverse learners.

5.4   Design Overview

Over the past three years, our research team has been developing a free online learning platform called TunePad that combines musical expression with Python computer programming (Fig. 5.1). Over this time we have refined successive prototypes with over 500 middle school and high school students in a variety of learning spaces including schools, libraries, summer camps, and other out-of-school programs. This work culminated in a public beta release in late 2018. Our hope is that this platform might serve as a foundation for prolonged interest, learning, and creative expression

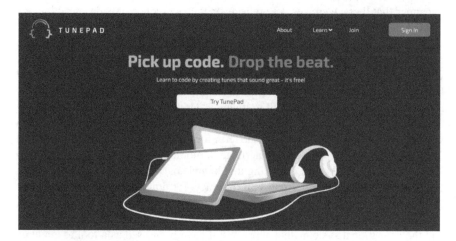

Fig. 5.1 TunePad is a free, online platform with the goal of empowering diverse learners to create and share music through code

for young people whose participation in computing has been historically marginalized. TunePad is part of a broader effort in collaboration with the EarSketch (Freeman et al. 2019; Magerko et al. 2016) team at the Georgia Institute of Technology to create an ecosystem of music + coding learning opportunities that span a range of ages, learning settings, and forms of musical expression.

There have been many environments designed to support learning at the intersection of music and code, going back to the early work of Bamberger (Bamberger and diSessa 2003; Bamberger 2013) and now including projects such as EarSketch (Freeman et al. 2019; Magerko et al. 2016), Sonic Pi (Aaron and Blackwell 2013), and Jython Music (Manaris et al. 2016). There are also numerous general-purpose learning environments such as Logo (Papert 1980), Scratch (Resnick et al. 2009), and Pencilcode (Bau et al. 2015) that support musical creation at various levels of sophistication. In creating TunePad, we draw inspiration from these projects, but we also feel that there is something important missing in the options available to learners and educators. Inspired by commercial apps like GarageBand, we designed TunePad with the goal of being much more playful and exploratory, while at the same time providing capabilities for serious musical creation.

When someone first visits the TunePad site, they can browse through a library of popular songs (Fig. 5.2), follow step-by-step interactive tutorials, or listen to music created by other users. All of the music on TunePad is built entirely with Python code and can be remixed as a starting point for new projects. Learners can share their work on the TunePad site or through existing social media platforms.

Fig. 5.2 When users first visit TunePad, they can browse through a library of popular music or follow interactive tutorials. All of the music on TunePad is built with Python code, and anything can be remixed as a starting point for new compositions

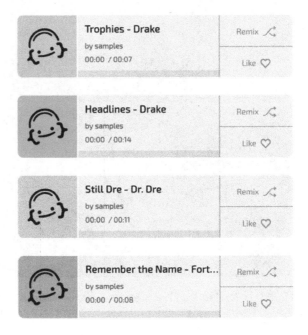

TunePad adopts a computational notebook paradigm that has been popularized by tools such as Jupyter (Kluyver et al. 2016) and Mathematica, both widely used in data science communities. TunePad projects take the form of interactive web pages called Dropbooks that combine playable musical instruments, text, lyrics, multimedia elements, and runnable segments of Python code. The term Dropbook combines the phrase "computational notebook" with the idea of dropping an album or dropping the beat. As learners develop their projects, they can add new cells containing text, multimedia, and instruments (drums, bass, synths, and so on). For example, adding drums reveals a drum pad that can be played directly with a mouse, keyboard, or touchscreen (or attached MIDI device). Instruments also have a number of different voices to provide a more interesting array of sounds to explore. To "record" tracks, users write short Python scripts that can also play the instruments. Figure 5.3 shows a script that plays a simple run of hi-hats containing random stutter steps (a common pattern in popular music).

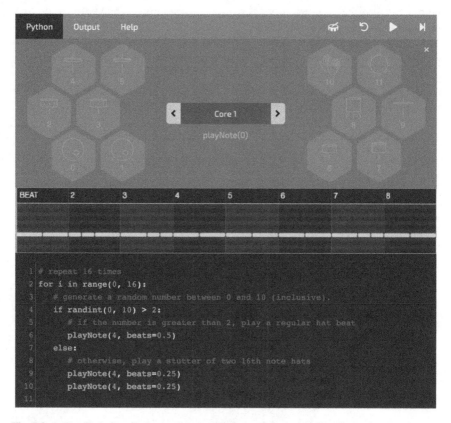

Fig. 5.3 In TunePad, short Python scripts are tightly integrated with playable musical instruments (top) and interactive piano rolls (middle). These three elements form cells that can be embedded with other cells in interactive web pages called Dropbooks

This code includes a for-loop (that sets the basic repeated pattern) and a conditional statement that uses a random number to decide whether to play a single 8th note or two 16th notes. With only 10 lines of code, the program conveys underlying meanings and patterns that can be more difficult to see using traditional musical notation systems. Specifically, there is a foundational pattern that should be punctuated with seemingly random stutter steps that occur roughly two times out of ten. It's also easy to change the code to experiment with musical alternatives. What if we increase the frequency of stutter steps, or change the sound played, or speed up the stutter steps to 32nd notes instead of 16th notes?

Figure 5.3 gives a small example of the rich potential that exists at the intersection of music and code. Our curricular materials explore core concepts such as loops, variables, functions, parameters, and conditionals. What was striking for us was how naturally these computing concepts flowed from the task of reproducing popular genres of music. Another example is a function that gets used in many of our introductory activities to create a "drum machine" in TunePad. The function takes a string representing a pattern of notes to be played and a number representing the percussive sound.

```
def beatMachine(sound, pattern):
    moveTo(0)  # move to the beginning of the measure (beat 0)
    for x in pattern:
        if x == '-':
            rest(0.5)
        else:
            playNote(sound, beats=0.5)
```

In a short amount of code, the function illustrates the idea of looping through a list of elements, the use of conditionals to play sounds or rests depending on the symbol in the pattern parameter, the use of functions and parameters, and the idea of modularity and code reuse.

5.5 Summer Camps

We have been working on the design and development of TunePad in collaboration with a number of partner organizations (including schools, libraries, community centers, and youth programs). In total, we have helped run over 20 events (ranging from around 60 min to 3-weeks in duration). More recently, we have begun formal data collection with IRB approval. Here we share data from two out-of-school camps with middle school learners conducted in the summer of 2019. Both camps emphasized technology and coding at the intersection of art, music, and performance. In collaboration with our partner organization, our team participated in these events with daily sessions on music and coding, involving around an hour of focused work with

Fig. 5.4 Participants at one of the camp's showcase performances

TunePad. A typical day would include some form of independent warm-up activity, a 15–20 min structured overview of core concepts from music and/or coding, and time for students to work individually or in pairs to practice applying the concepts. Each camp was three weeks long and featured musical performances and showcases as culminating events for parents, friends, peers, and families (Fig. 5.4).

For this chapter, we share data from observations of student work and learner-created projects and artifacts. Before presenting the results, it is important to note that TunePad was only one part of larger programmatic activities that involved learning about the history of hip hop and social justice and engaging in making and tinkering activities. We used these case studies to reflect on music and computing as intersecting literacies.

5.6   Learner Case Studies

Kala   Ten-year-old Kala (all names are pseudonyms) was inspired by music. In one of our workshops, we formed a drum circle that served as a hands-on introduction to making music with code. During the drum circle participants played percussion instruments and were introduced to specific symbols (#, -, !, *) related to patterns and dynamics that were embedded in a drum machine simulation coded in TunePad. The project was preloaded onto their laptops and they were instructed to manipulate the patterns and dynamic symbols to create their own rhythmic compositions after the drum circle ended. When they were ready participants shared what they created with the group over a loudspeaker. Kala shared her composition, which was good but did not exactly fill up the 2 musical measures in 4/4 time; consequently, there was an awkward silence at the end before her pattern looped. She was praised for her work, but she took it upon herself to go back and fix her composition so that it filled the 2 bar measure at which point she re-shared her work. As part of another project, Kala was determined to translate a song that she loved into Python code. Kala found a piano tutorial on YouTube that showed what notes to play for the actual pitches on the TunePad keyboard. Up to this point, Kala had simply created individual playNote commands, one after another, to transcribe the melody from the YouTube video without recognizing repeated patterns and phrases. We think that this is an important phase of learning that needs to play out in its own time. Working slowly and inefficiently can help set the stage for further learning opportunities where concepts such as loops or functions will make intuitive sense because their usefulness is highlighted against a backdrop of hard-earned personal experience. One advantage of informal learning activities separated out from more structured learning goals is the freedom to explore ideas at learners' own pace. Kala put hours of work into the project for presentation at the camp showcase. Her composition ultimately included several synchronized tracks with piano and drum components.

Serena Like Kala, Serena was motivated to create music. For one activity, we asked participants to compose an original song inspired by a visual image. For the project, there was a specific rhythm that Serena was able to clap out that she wanted to include in her composition. One difficulty she had with her rhythm is that TunePad only supported a 4/4 time signature at that time, which didn't match her vision for the song. This presented an unusual problem when she tried to sync her tracks because there was an eighth note gap that she didn't want. She asked for help from a facilitator with the problem. After the facilitator asked her a few questions, she took the opportunity to ask one herself: "Don't we have to use math to solve it?" A smile came over her face, and they began to approach the problem mathematically. First, they had to determine the time signature of the beat, which turned out to be seven-eighths time (seven eighth notes per measure to be repeated). Then she had to determine how many bars it would take to fill up the time while avoiding unwanted gaps. After trial and error, Serena decided that repeating her pattern eight times would give her 56 8th notes, which is a multiple of both 7 and 8, allowing her 7/8 time signature to line up with the 4/4 built-in time signature of TunePad.

Raphael Raphael is a music student who studies piano. This stronger musical background enabled Raphael to engage in more in-depth explorations of musical composition with TunePad. For one of his projects he composed a melody at home that he committed to memory. Working at camp, he then used the piano instrument to code the melody with Python, something he seemed proud of. As he got deeper into the material he composed a remarkable three-part brass composition that included elements like call and response with a counterpoint melodic feel. Raphael also experimented with transposing arpeggios. An arpeggio is an excellent example of an algorithmic structure in music. It consists of a pattern of notes that are played in rapid succession up or down a scale from a base note position. This technique is commonly used in digital music as a way to punctuate harmony with a rapid rhythmic feel. To implement an arpeggio in TunePad typically involves a loop over a pattern of notes represented as a list data type in Python. Although Raphael's code did not make the most elegant use of loops to capture repeated phrases, he obviously had a handle on how to use the code to create the effects that he wanted with some trial and error. As with Kala, we felt that it was much more important for learners to have ownership and independence in their coding than for their programs to be perfect. This is more in line with common practices in informal learning environments where adult facilitators ideally follow the lead of youth and provide just-in-time support for learners as they are ready for more information and ask for help.

5.7 Discussion

Looking at each of these three cases of student work in music and coding, we saw some common themes of creative energy, a motivation to create (or recreate) musical compositions, and an inventiveness and tenacity when it came to seeing their vision realized. For example, after Kala discovered the YouTube video, she painstakingly transcribed the piano roll in the video into code. At varying levels, we saw learners building on their musical knowledge to anchor their learning of computer programming. This was perhaps most obvious in Raphael's arpeggios, but was also evident in Kala and Serena's projects as well. Finally, there was a freedom to learn code on their own terms without having a correct or elegant solution foisted on them by facilitating adults. These three brief excerpts of student experiences with TunePad in our summer camps give us a slightly more concrete way to reflect on the development of computation as a literacy. Returning to the implications of a literacy framing from earlier in the chapter, our ongoing work with students has shaped our thinking in a number of ways. First, culturally responsive CS education programs that understand coding as a kind of literacy might find it productive to consider the possibility of multiple forms of literacy—dominant forms prescribed by the academy and teaching standards might not be the only way to express ideas through code. Or, as Kafai puts it: "Programming that prizes coding accuracy and efficiency as signifiers of success is boring. To learn programming for the sake of programming goes nowhere for children unless they can put those skills to use in a meaningful way" (Kafai 2016). We see this as especially relevant in the domain of music. It's entirely plausible that the ways we use code to create music won't be consistent with the ideals and norms established by formal CS education. Here, the examples from Kala and Raphael show us how learners can use computational tools to create music that don't necessarily align with the broader goals of Computer Science.

Second, music is also a kind of literacy. Looking beyond those with formal training in musical notation systems, everyone has a degree of fluency in musical forms and conventions that have been developed since infancy through participation in human culture. Even without formal music education, we all have some ability to express musical ideas (even as simple as clapping out a rhythm) (Bamberger 2013). All three of our learners worked hard to translate intuitive musical ideas into functioning projects, each drawing on different resources to realize their vision. Kala found a YouTube video that she was able to play, pause, and rewind, over and over again to transcribe her song note by note, while Raphael was able to draw on more formal musical training (a more advanced stage of musical literacy). From these examples, we have started to think of learning music and learning programming as parallel endeavors that, counter-intuitively, may be easier to tackle together than to learn in isolation—a paradox of multiple literacies. To give an intuitive sense for why this might be the case, learning to code is hard, but music provides a motivating, authentic context in which concepts make sense and are *useful* in a direct and meaningful way (Freeman et al. 2019). Learning music is also hard, but coding offers a productive *restructuration* (Wilensky and Papert 2010) of music that might be more accessible to novices than traditional musical notation (Bamberger and diSessa 2003).

Third, in the domain of music, we think of coding as more of a humanity than an engineering discipline. Like poetry, dance, and literature, engaging in the craft of computer programming can enrich the human experience as a joyful and fulfilling activity in its own right. These activities become even more meaningful when they are directed toward creative expression and then shared with others (Kafai 2016). Each of the three examples shows learners who were driven to see their creative vision enacted in the form of a tangible musical artifact.

Finally, the TunePad project is interested in what happens when you bring musical, social, and political ideologies embodied in movements like Hip Hop into collision with more established forms of coding literacy and CS education. As the literacies of coding and music collide, can we recognize, value, and make space for diverse ways of knowing and understanding that learners bring with them when they first encounter programming? Can we be open to a reimagining of the dominant forms of coding literacy that might make us uncomfortable as CS educators? Can coding be transformed (and transformative) in ways that might lead us to a more diverse computational future?

Acknowledgements The authors would like to thank Brian Magerko and Jason Freeman from Georgia Institute of Technology, and Nichole Pinkard and Amy Pratt from Northwestern University. We also gratefully acknowledge Cortez Watson, Brian Andrus, Izaiah Wallace, and all of the participants in our summer camp. This research was supported by grants DRL-1612619, DRL-1837661, and DRL-1451762 from the National Science Foundation. Any opinions, findings, and conclusions or recommendations expressed in this material are those of the authors and do not necessarily reflect the views of the NSF.

References

Aaron S., & Blackwell, A. F. (2013). From Sonic Pi to Overtone: creative musical experiences with domain-specific and functional languages. In *Proceedings of the First ACM SIGPLAN Workshop on Functional Art, Music, Modeling & Design* (pp. 35–46). ACM.

Abbott, B. P., Abbott, R., Abbott, T. D., Abernathy, M. R., Acernese, F., Ackley, K., et al. (2016). Observation of gravitational waves from a binary black hole merger. *Physical Review Letters, 116*(6), 061102.

Bamberger, J. (2013). *Discovering the musical mind: A view of creativity as learning.* Oxford University Press.

Bamberger, J., & diSessa, A. (2003). Music as Embodied Mathematics: A study of mutually informing affinity. *International Journal of Computers for Mathematical Learning, 8,* 123–160.

Bau, D., Dawson, M., & Bau, A. (2015). Using pencil code to bridge the gap between visual and text-based coding. In *Proceedings of the 46th ACM Technical Symposium on Computer Science Education* (p. 706). ACM.

Brennan, K., & Resnick, M. (2012, April). New frameworks for studying and assessing the development of computational thinking. In *Proceedings of the 2012 Annual Meeting of the American Educational Research Association* (Vol. 1, p. 25). Vancouver, Canada.

Chiang, T. (2019). *Exhalation.* Pan Macmillan.

diSessa, A. A. (2018). Computational literacy and "the big picture" concerning computers in mathematics education. *Mathematical Thinking and Learning, 20*(1), 3–31.

Event Horizon Telescope Collaboration. (2019). First M87 event horizon telescope results. I. *The shadow of the supermassive black hole.* arXiv preprint arXiv:1906.11238.

Freeman, J., Magerko, B., Edwards, D., Mcklin, T., Lee, T., & Moore, R. (2019). EarSketch: Engaging broad populations in computing through music. *Communication ACM 62*(9), 78–85.

Grover, S., & Pea, R. (2013). Computational thinking in K–12: A review of the state of the field. *Educational Researcher, 42*(1), 38–43.

International Human Genome Sequencing Consortium. (2001). Initial sequencing and analysis of the human genome. *Nature, 409*(6822), 860.

Kafai, Y. B. (2016). From computational thinking to computational participation in K–12 education. *Communication ACM, 59*(8), 26–27.

Kafai, Y. B., & Burke, Q. (2013, March). The social turn in K-12 programming: moving from computational thinking to computational participation. In Proceeding of the 44th ACM technical symposium on computer science education (pp. 603–608).

Kluyver, T., Ragan-Kelley, B., Pérez, F., Granger, B, Bussonnier, M., Frederic, J et al. (2016). Positioning and power in Academic Publishing: Players, agents, and agendas. *Chapter Jupyter Notebooks—A publishing format for reproducible computational workflows* (pp. 87–90). IOS Press

Kohn, W. (2003). Nobel lectures, chemistry 1996–2000 (p. 213). World Scientific Publishing Co, Singapore.

Magerko, B., Freeman, J., Mcklin, T, Reilly, M., Livingston, E., Mccoid, S., & Crews-Brown, A. (2016). Earsketch: A steam-based approach for underrepresented populations in high school computer science education. *ACM Transactions on Computing Education (TOCE), 16*(4), 14.

Manaris, B, Stevens, B, Brown, A. R. (2016). JythonMusic: An environment for teaching algorithmic music composition, dynamic coding and musical performativity. *Journal of Music, Technology & Education, 9*(1), 33–56.

Margolis, J. (2010). *Stuck in the shallow end: Education, race, and computing.* MIT Press.

Papert, S. (1980). *Mindstorms: Children, computers, and powerful ideas.* Basic Books, Inc.

Pople, J. (2003). Nobel lectures, chemistry 1996–2000 (p. 246). World Scientific Publishing Co, Singapore.

Resnick, M., Maloney, J., Monroy-Hernández, A., Rusk, N., Eastmond, E., Brennan, K., et al. (2009). Scratch: Programming for all. *Communication ACM, 52*(11), 60–67.

Shute, V. J., Sun, C., & Asbell-Clarke, J. (2017). Demystifying computational thinking. *Educational Research Review, 22*(2017), 142–158.

Vee, A. (2017). *Coding literacy: How computer programming is changing writing.* MIT Press.

Weintrop, D., Beheshti, E., Horn, M., Orton, K., Jona, K., Trouille, L., & Wilensky, U. (2016). Defining computational thinking for mathematics and science classrooms. *Journal of Science Education and Technology, 25*(1), 127–147.

Wilensky, U, & Papert, S. (2010). Restructurations: Reformulations of knowledge disciplines through new representational forms. *Constructionism 2010.*

Wing, J. M. (2006). Computational thinking. *Communication ACM, 49*(3), 33–35.

Zweben, S., & Bizot, B. (2018). 2017 CRA Taulbee Survey. *Computing Research News, 30*(5), 1–47.

Michael S. Horn is an Associate Professor of Computer Science and Learning Sciences at Northwestern University where he directs the Tangible Interaction Design and Learning (TIDAL) Lab. His research involves the creation of playful technology-based learning experience to support computational literacy.

Amartya Banerjee is a research associate in Computer Science at Northwestern University.

Melanie West Melanie West is a Ph.D. student in Learning Sciences at Northwestern University and the Director of Curriculum for the TunePad project.

Chapter 6
Programming in Primary Schools: Teaching on the Edge of Formal and Non-formal Learning

Katharina Geldreich and Peter Hubwieser

Abstract While several countries have already introduced Computer Science or programming into their primary school curricula (e.g., the UK, Australia, or Finland), Germany has not yet developed mandatory guidelines on how to deal with these matters. Although there is an agreement that students of all ages should gain insight into the recognition and formulation of algorithms, the focus in primary school is often still on the mere use of computers. Programming courses, on the other hand, are increasingly found in extracurricular activities. It is still open to what extent and in what form algorithms and programming can and should be introduced in primary schools in the longer term. To help answer this question, we trained 40 primary school teachers in algorithms and programming and examined how they implement the topics in their individual schools. Among these are teachers who teach programming in class (formal learning) as well as teachers who offer their students extracurricular programming activities on a voluntary basis (non-formal learning). We interviewed all teachers about how they implemented the topics, what advantages they saw in the individual formats, and what challenges they encountered. In this paper, we outline our didactical approach as well as the results of our interview study.

Keywords Programming · Primary school · Teacher training · Interviews

6.1 Introduction

In recent years, the discussion about the necessity of Computer Science (CS) and especially programming in primary education has grown steadily (Webb et al. 2017; Bell and Duncan 2018). The early development of key understanding, skills, and thinking approaches emerging from CS seems to have several positive effects on

K. Geldreich (✉) · P. Hubwieser
School of Education, Technical University of Munich, Munich, Germany
e-mail: katharina.geldreich@tum.de

P. Hubwieser
e-mail: peter.hubwieser@tum.de

© Springer Nature Singapore Pte Ltd. 2020
M. Giannakos (ed.), *Non-Formal and Informal Science Learning
in the ICT Era*, Lecture Notes in Educational Technology,
https://doi.org/10.1007/978-981-15-6747-6_6

children. Learning to use computers not only as users but also as creators and gaining positive experiences in computing can strengthen their self-confidence in CS and technology in general (Duncan et al. 2014; Topi 2015). It may also prevent common misconceptions and prejudices toward CS regarding the nature of the subject and the role of gender (Moorman and Johnson 2003; Engeser et al. 2008; Funke et al. 2016a). In addition, computational thinking—which is generally defined as the mental activity of abstracting problems and formulating automatable solutions (Wing 2006)—has the potential to improve students' problem-solving skills in other subjects as well (Yadav et al. 2014).

Several countries have already included aspects of CS in their primary school curricula, e.g., Australia (Falkner et al. 2014), Finland (Kwon and Schroderus 2017), the UK (Brown et al. 2013), and Switzerland (D-EDK 2016). Apart from these formal learning settings, there are numerous non-formal offerings aimed at promoting children's interest in and knowledge of CS. They offer the opportunity to deal with CS, even if it is not part of the curriculum. Many of them focus on programming or coding, such as the website *code.org*[1] or the programming clubs *Code Club*[2] and *Coder Dojo.*[3] These extracurricular activities are voluntary and can provide experiences that are not anchored in the curriculum or not possible in regular classroom settings (Lunenburg 2010).

It is still unclear to what extent and in what form CS and programming can and should be introduced in primary education in the longer term, and what role extracurricular offers should play in this context. To help answer these questions, we wanted to include the opinions and experiences of primary school practitioners. We trained 40 primary school teachers in algorithms and programming and examined how they implement the topics in their schools. We did not specify the setting of this implementation—both formal and non-formal formats were possible. In the course of a school year, we conducted exploratory interviews with all teachers. The focus was set on the following research questions:

- Which are the most common settings for introducing algorithms and programming?
- What advantages do the teachers see in the respective settings?
- What challenges and limitations do the teachers encounter?

In this article, we first give a brief introduction to CS in primary education and extracurricular offerings on CS. We also give an insight into the teaching concept the teachers got to know as part of their teacher training. Afterwards, we will describe the research design and methods of our study as well as the results of the interviews. After discussing the results, we will give an outlook on our future research.

---

[1]https://code.org/.

[2]https://codeclub.org/.

[3]https://coderdojo.com/.

6.2 Background and Related Work

One can see an increasing consensus in CS Education that beginning to learn CS in primary school is not only possible but also beneficial for learning as well as developing self-esteem and motivation (Webb et al. 2017). Besides, research is being conducted into how computational thinking and CS can and should be integrated into other subject matters (Yadav et al. 2014; Weng and Wong 2017; Friend et al. 2018).

Although Germany has not yet developed binding guidelines for dealing with the topics, the relevance of CS in primary school is becoming increasingly evident. In its strategy paper on education in the digital world, the German KMK[4] states that competencies on *recognizing and formulating algorithms* should be included in the curricula of all school types (KMK 2017). The German Informatics Society (GI) goes even further and formulates competencies in five different content areas that students should develop during primary school (Best et al. 2019). There are also various research efforts focusing on how we can allow children to acquire basic knowledge in the field of CS and which methods and contents are suitable for German primary schools (Diethelm and Schaumburg 2016; Gärtig-Daugs et al. 2016; Geldreich et al. 2016; Bergner et al. 2017; Goecke and Stiller 2018; Magenheim et al. 2018).

However, there is only little work on German primary school teacher's beliefs and opinions on CS. Funke et al. (2016b) conducted an interview study with six primary school teachers without any previous experience in CS. In this study, they conclude that the interviewed teachers have no concrete picture of CS in primary school but do have some beneficial preconceptions and attitudes. Best (2019) conducted semi-structured interviews with eleven primary school teachers without any relevant prior knowledge about their views on CS as a discipline and subject in primary school. He repeated the interviews with three teachers after they had gained first teaching experience with Bee-Bots.[5] The findings show that the teachers consider CS education to be important for primary school, but also for the lives and future careers of the students. The opinions where this education should take place were heterogeneous: as an independent subject, integrated into several subjects, integrated into one subject, or as an extracurricular activity. They assumed that boys have a higher interest in CS than girls and are convinced that this must be counteracted already in primary education.

There are further international studies that focus on teachers' experiences and perspectives. Sentance et al. (2017) interviewed 15 teachers about their use and experience of the micro:bit,[6] a physical computing device. They categorize different approaches and instructional styles to teaching with physical computing and identify teachers who can be classified as either *inspirers, providers,* or *consumers.* Black et al. (2013) conducted a questionnaire-based study on teachers' perceptions of how to make computing interesting for students. Out of 115 responses from British CS

---

[4]Kultusministerkonferenz (literally "conference of ministers of education") is the assembly of ministers of education of the German states.

[5]https://www.tts-group.co.uk/bee-bot-programmable-floor-robot/1015268.html.

[6]https://microbit.org/.

teachers, several factors were identified as most important for engaging students. Based on the results, they give specific recommendations where teachers should be supported in this matter. Yadav et al. (2016) examine the experiences and challenges that novice CS high school teachers face in the classroom. They conducted 24 semi-structured interviews and identified several challenges, including isolation, lack of adequate CS background, and limited professional development resources. Duncan et al. (2017) analyzed the feedback of 13 teachers participating in a study that examines the implementation of new primary school topics based on computational thinking in New Zealand. The teachers had no previous experience in teaching CS and volunteered to take part in a program where they receive professional development and support to integrate computational thinking and CS in their teaching. They were asked to complete a feedback form each time they taught a session that focused on CS or computational thinking. Based on these feedback forms, they identified ways in which the teachers could integrate computational thinking into their current teaching, the key concepts they were able to engage students with, and their confidence in delivering the material.

6.3 Context

This study is part of a two-year project called *AlgoKids—Algorithmen für Kinder (in English: "Algorithms for Children")*, which is funded by the Bavarian Ministry of Education. The project investigates how primary school teachers can be prepared and supported to teach the topics *algorithms* and *programming* in Bavarian primary schools. In addition, both the implementations and experiences of the teachers are scientifically analyzed and evaluated. In two multi-day professional development trainings, the participating teachers received the opportunity to expand their computing knowledge (Geldreich et al. 2018). After the training, they were provided with additional online material as well as the possibility to seek further support if required.

The project is based on an already field-tested and evaluated programming course for primary school, which is aimed at third and fourth-grade students (Geldreich et al. 2019). We have tested it in practice with whole school classes as well as an extracurricular activity with children who have participated voluntarily. During the project, the teachers implemented the course at their school. They have not been told in which subject context this should take place and whether they should approach the topics in a formal or non-formal setting.

The course includes unplugged activities as well as working with the visual programming language Scratch (Maloney et al. 2010). At the end of the course, the students should understand that a device is following an algorithm that is implemented by programming the device. They should also get familiar with the process of testing and debugging a program and get to know the basic algorithmic structures *sequence, selection,* and *iteration.* At the same time, the course promotes the computational thinking skills of *algorithmic thinking* (e.g., follow algorithms,

Fig. 6.1 Pictorial representation of making a sandwich

create algorithms to solve problems), *decomposition* (breaking down problems into smaller steps), *logical reasoning* and *evaluation* (e.g., identifying possible solutions and choosing the best one) (Berry 2015). The course concept is described in the following.

6.3.1 What is an Algorithm

Since most of the students do not have any prior knowledge in programming or CS in general, the first step is to give them a basic idea of how computer programs work. They initially work unplugged, i.e., without a computer, and program in everyday language. In the first step, they program the teacher—she or he plays a robot and is supposed to perform small tasks in the classroom, such as opening the window. Since the teacher only follows particular commands, the children quickly realize that each step in an algorithm must be formulated in an understandable, precise, and unambiguous way. Larger actions must be broken down into sub-steps. It is also explored where they encounter algorithms in their everyday lives, for example, in the form of handicraft instructions or recipes. In different tasks the students practice describing sequences in natural language, for example, they convert a pictorial instruction into unambiguous language-based commands (see Fig. 6.1).

6.3.2 Programming Unplugged

Subsequently, the description of algorithms is further explored. The students use everyday language, symbols, and haptic Scratch blocks to program each other and solve different tasks (Fig. 6.2, left). As soon as a task has been solved, the solution can be executed in a grid and checked for mistakes (Fig. 6.2, right). This way, they can physically experience what later happens in a programming environment. We have designed the tasks in a way that allows them to be solved by using *selections* and *iterations*, but also by *sequences*.

Fig. 6.2 Task (left) and corresponding grid in the classroom (right)

6.3.3 Programming in Scratch

After these unplugged exercises, the students are introduced to the programming environment Scratch. To enable the students to concentrate on using the Scratch programming environment, they first work on some tasks they already have solved unplugged. Next, they work on a learning circle in which the core operations of Scratch are gradually introduced and which the students can master at their own pace (Fig. 6.3). Starting from questions regarding software handling, the stations lead from simple *sequences* to the implementation of *selections* and *iterations*.

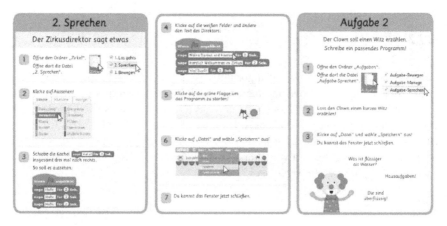

Fig. 6.3 Station of the Scratch learning circle

6.3.4 Planning Programs

The last step in the teaching concept leads the students to plan and implement their own program ideas. In individual or partner work, the students think up their own Scratch story, write it down in a script (Fig. 6.4), and implement it in Scratch. To get comparable results, we set the following mandatory requirements for the students' projects. The programs should (1) work on more than one sprite (2) move the sprites during execution (3) comprise at least one loop, and (4) include at least one conditional statement. After meeting these requirements, the students could continue their programming work without any further guidelines. The children present their programs in front of the class and are given the opportunity to comment on their projects.

Fig. 6.4 Project script

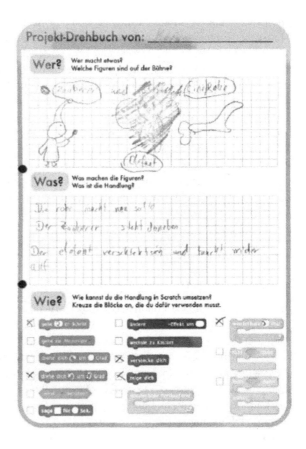

Table 6.1 Age distribution of participating teachers

Age	Number of teachers
Under 30 years	8
30–40 years	16
41–50 years	9
Older than 50 years	7
Total	40

6.4 Methods

6.4.1 Participants

The twenty schools which participate in the project were selected by the Bavarian Ministry of Education. In order to reflect the Bavarian school landscape, they chose primary schools from all government districts. To make a comprehensive selection, they also took into account the size, experience in digital education, and technical equipment of the schools. In total, we worked with 40 teachers—two from each primary school (two males, 38 females). The age of the participants ranged from under 30 years to over 50 years, while the group of 30–40-yearolds made up the largest part (see Table 6.1). Across both groups, 27 teachers had no previous experience in CS at all, 13 teachers had CS for 1–3 years as an elective or compulsory subject in school. We also assessed whether the teachers participated on their initiative or the initiative of their principal, or whether the initiative was evenly divided between the two. The answers were distributed almost equally among the three possible options.

6.4.2 Data Collection

To get insights into the implementations and experiences of the teachers that took part in the project, we conducted exploratory interviews. The exploratory interview is not—like the classical interview—an asymmetrical form of communication. Although there is still a separation of roles between the *interviewer* and the *interviewee*, the interview situation is a quasi-normal conversation (Honer 2011). The exploratory interview does not follow any specific rules, the questions, however, should be asked as openly as possible. Nevertheless, the interviewer always has the possibility to follow up on interesting points or to steer the conversation in a certain direction with suitable questions (Ullrich 2006).

With few exceptions, the interviews in our study were conducted jointly with both teachers of each school. They were led by one researcher who has already given the teacher training and who knew the teachers well. They were asked to tell what they have done with the students so far and in which context they introduced *algorithms* and *programming*. In the course of the interviews, it was also discussed what learning gains they had observed among their students, what challenges they encountered, and whether they were able to identify differences between boys and girls. The interviews were audio-recorded and transcribed. In the following, we present selected results from the exploratory interviews that relate to the setting the teachers introduced *algorithms* and *programming*. We conducted a total of 19 interviews, which lasted between 30 min and two hours.

6.4.3 Data Analysis

The data were analyzed within the qualitative data analysis software MAXQDA. Based on our research questions, we first categorized the transcripts regarding two main categories: formal and non-formal learning settings. Following grounded theory, we then started with open coding by attaching codes to the teacher statements (Corbin and Strauss 1990). In an inductive process, we searched for emerging patterns by grouping codes from both main categories (Glaser in Walsh et al. 2015). The overall objective at this point was to create themes that should lead to a structure for reporting our results.

6.5 Results

Although the teachers were provided with all resources from our teaching approach, they were free to modify or expand the materials or develop their own learning materials and scenarios. Even if all teachers have followed our teaching concept in general, there was considerable variation in the specific setting of the implementations and their experiences. Teachers from fourteen schools implemented programming exclusively in a formal setting, in three schools they offered programming clubs in a non-formal setting. Two schools collected experience in both settings. We report on the data in relation to five areas that emerged from the analysis—all areas contain results that refer to both formal (F) and non-formal (N-F) implementations:

- Implementation in school
- Student engagement
- Teachers' confidence
- Challenges and concerns
- Gender issues.

All interviews were conducted in the German language. The anchor examples below were translated into English by the authors.

6.5.1 Implementation in School

Both the teachers who implemented programming in regular lessons and those who offered it as an extracurricular club considered it a useful activity for the students:

> I think, on the one hand, it's very motivating, it's modern, it's simply a medium that children have to deal with in a meaningful way. On the other hand, with all these unplugged modules beforehand, we don't just place them in front of computers and let them do whatever they want. The precise formulation, bundling an idea and implementing it within Scratch as a program - this is highly complex. (F)

> I think it helps the students to think in a more structured way. They have to make a plan in their heads – they can try things out, but they also have to think about it carefully. That's often not the case in regular lessons. (F)

For many teachers, it is a challenge that programming is not anchored in the curriculum. They would like to have more freedom in the timetable to allow them to implement such topics more flexibly. At the same time, however, many think that it could be problematic for a lot of teachers if it were required in future curricula:

> It's a pity that it's not in the curriculum because there is so much potential in the children. You could really tap into that. They are so motivated and have no inhibitions and fears. (F)

> On the one hand, it should be anchored in the curriculum, otherwise, nobody will do it. On the other hand, I also find it difficult to institutionalize it - how do you want to assess the performance of the students? (F)

> I believe that interested teachers implement it, whether or not it's part of the curriculum. But many teachers have no affinity for it. And I don't think they would do it even if it was in the curriculum. (F)

> I believe that programming could become a new cultural technique in the foreseeable future and that everyone should get insight. But the place for it in primary school has yet to be created. Finding a place in regular classes is difficult. (F)

Some teachers have opted for an extracurricular offer because they cannot provide enough time for programming during regular lessons. In addition, it was mentioned that only those children who are really interested in the topic sign up for a club:

> We have outsourced programming into a club. It would be difficult for us to implement it in everyday school life. (N-F)

> If you offer a programming club, you would always have a designated time for that. And you have children who are really interested in it. (F)

> There were children in the club who made a conscious decision to participate. They find it cool to learn more about Computer Science. (N-F)

The majority of the teachers are in favor of programming being included in the curriculum of primary schools. However, there is disagreement about the context in which this should or could happen:

> In mathematics, you could include sessions about giving precise instructions – because mathematics works similarly. You must follow a certain sequence of commands or rules when you do a calculation. German lessons would also be possible – they could write a recipe or other instructions. (N-F)

It's something interdisciplinary. It has something of mathematics, of language, of everything. In the curriculum there is the area "media education", but it is very vague and easy to avoid. It would have to be made much clearer in the curriculum how something like this can be linked with the other subjects. (F)

It is also nice when an expert comes from outside and offers an activity for the children. But that's always this one special project day – and it shouldn't be like that. You could do a lot during regular lessons. (N-F)

Although all teachers considered the unplugged activities in our teaching concept to be necessary and have had positive experiences with them, they find it important to program on the computer as well:

I don't quite understand the idea of only doing the preliminary work for programming and to program unplugged exclusively. Of course, you can build understanding for the algorithmic structures – but isn't it like coffee without milk? (F)

6.5.2 Student Engagement

A recurring theme in the interviews—whether programming was implemented in a formal or non-formal setting—was the emphasis on the students' enjoyment of the sessions and the high level of engagement they demonstrated. Several teachers pointed out, that they were surprised about the engagement of individual students:

It is so nice when the students leave and say: "Wow, that was such an awesome lesson today!" They have such great achievements. (N-F)

All the students were very interested. Some children, who are otherwise very reserved, suddenly became really active. (F)

I can see that children in the club are developing real enthusiasm. They've already bought Scratch books, registered in the online community and share their projects there. They even told me their older brothers and sisters started programming because they told them about it. (N-F)

There were several comments from teachers who implemented programming in a non-formal setting where they stated that all students should have the experience of learning to program:

The motivation lasted the whole school year. If the club is canceled for any reason, the students asked me in the schoolyard: "Why is there no programming this week?" I would have wished that more students could have joined the programming club. (N-F)

Several teachers started computing with the entire class and later thought about diving deeper with students that showed the most interest:

We programmed half a school year with the entire class. Then, we thought about offering a club in the second half of the year. We have a lot of children who are really interested and could explore it in depth. (F)

Some teachers who have introduced programming in regular classes have expressed concerns about the seriousness of the activity or whether students are learning what was intended:

Everything was very simple and playful. I don't think they've realized yet that this is Computer Science – programming is a lot of fun for them. (F)

On the next level, I want programming to become a little more serious. It's not just about coding funny things – I want them to think about how to program specific actions. But at the same time, I don't want to slow them down. They are so full of joy and imagination. (F)

It must have added value. For sure, it's good for motivation. They enjoy programming a lot. There is a benefit in that because if they enjoy coming to class, they learn something. But do they always learn what they are supposed to learn? (F)

6.5.3   Teachers' Confidence

Many teachers were worried they would not be able to answer all the questions of the students. Some teachers first tried out certain contents and methods with a few students and only then ventured into a larger group:

At first, I tried out some exercises and methods with a few children from my class. We went to the computer room once a week for two months. After that, I felt comfortable to run the programming club on my own. Also, because I knew that I had your concept and material which I could stick to. A lot of things grew out of that. (N-F)

Some teachers noticed that they adopted a different teacher role than usual when programming with their students and felt quite comfortable with that. Despite some initial concerns, many even saw an advantage in not always knowing all the answers:

The role of the teacher is as it should be in exploratory learning. One can approach the individual children, respond to them, advise them. They decide what suits them best, think for themselves, become active and are not satisfied with ready-made solutions. They can bring in their ideas again and again. (F)

I was often clueless; stood by a student and had to admit that I had no idea. But that was also great because students realized that teachers aren't perfect either. And you grow together when you work on problems together. Sometimes the students came up with the solution – sometimes I came up with it. That was a great collaboration. (F)

Some teachers noted that before the project they were not at all interested in programming and now see it as a personal enrichment:

I am very grateful that I had the chance to participate in the project. It's so much fun and I've discovered hidden talents in myself. As a woman, I had the attitude that I wasn't interested in Computer Science. Well, I am now! (N-F)

Although some teachers have had positive experiences with programming as an extracurricular offer, they have reservations about programming with the whole class due to the high number of students:

Sometimes I wish there was a second person in the club with me. This person could help if the computer won't start or help the students when I'm busy. But the children are relaxed and know that sometimes it takes a while. They help each other a lot or just keep trying to solve the problem on their own. But in class, I have 29 children – that would be difficult to handle alone. (N-F)

6.5.4 Challenges and Concerns

The most frequent challenges for the teachers concerned the technical equipment of the school and not being able to respond adequately to all the needs and questions of the students:

> We were always two teachers when we programmed in class – that was OK. It would've been hard if I had been alone. The organization, these adversities with the equipment, that's all difficult. (F)

> Technical infrastructure and time are major problems. We don't have any system support at school and so I installed Scratch on all computers for a whole day. That's why it could fail – you save on staff and teachers are expected to do all the work voluntarily. (F)

For many teachers, it was a challenge to meet the different skills and knowledge levels of the students. Also, the use of a computer was a problem for many children:

> One student left the club after a while. He was already very advanced and had already programmed in C – his father is a computer scientist. The other students had never heard of programming. (N-F)

> I had students who already knew how to handle laptops, I had kids who knew Scratch and I had kids who never had any digital device in their hands. Balancing those differences was a big challenge in the beginning. (F)

> The handling of a computer is a big problem. How do I scroll down? How do I make a double-click? I had the feeling that many students couldn't get into the depth of programming because of this. (F)

When asked if they could imagine programming regularly with the whole class, they expressed conflicting concerns about the students' performance:

> I'm a little worried that at some point we'll reach a level where I can't help the students anymore. That gives me a bit of a stomachache because that doesn't happen to me in any other subject. You reach your limits at some point. That's not a problem with single programming sessions - but if we were to program a whole school year regularly. (F)

> I believe there are children who, even in the fourth grade, are not yet so far advanced in their cognitive abilities. They have simply already reached the maximum of their development with the other subjects in fourth grade. (F)

6.5.5 Gender Issues

When talking about differences between girls and boys, the teachers were positively surprised that girls were also interested in programming:

> I had already offered a computer club before. The girls didn't want to participate at all and said they were not capable of that. But with the programming club, it was different – many girls volunteered and wanted to take part. (N-F)

> Making positive experiences with Computer Science is important. I have noticed that many girls have discovered hidden abilities and got a sense of achievement — programming is not just for boys and isn't something they don't understand. (F)

Several teachers reported that the boys had more experience with computers and were more involved with them at home:

> I'd say the boys are better at handling the computer. Which is probably just because they have more contact with it at home. That doesn't mean they can do it better in general. But I think they just have more experience with it. Whereby girls have more patience when something doesn't work. (F)

> The boys in my club are the ones who are more involved with it at home. They sign up in the online community, download Scratch, get books and program at home. They approach me with specific project ideas they got at home and want to implement it in the club. I haven't heard that from the girls yet. (N-F)

6.6 Discussion

Returning to the research questions mentioned in Sect. 6.1, we first wanted to investigate in which setting the teachers in *AlgoKids* introduce the topics *algorithms* and *programming*. Out of a total of twenty schools, fourteen schools exclusively chose a formal setting during regular school days and three schools decided to offer an extracurricular programming club in an informal setting. Two schools decided to test both settings. It should be noted that in Bavaria the school administration must approve all extracurricular activities. These hours are then added to the teachers' working time. As there is currently a shortage of teachers at many primary schools, club lessons are often not approved.

As an advantage in favor of programming in a formal learning setting, it is mentioned that programming generally helps students to develop a more structured thinking and all children should be given this opportunity. At the same time, it could be an opportunity to reduce the gender gap regarding the students' interest in CS and the abilities in using the computer. Individual statements show that the family home can have a great influence on this previous knowledge. To ensure social justice, one would have the chance to take countermeasures in class. Another advantage of a formal setting was initially perceived as worrying by some teachers—the changing teacher role. However, after gaining initial experience, teachers reported that they enjoyed the changing role and were even able to build a better connection with their students.

The missing legitimacy in the primary school curriculum and the associated lack of time is primarily cited as a challenge for programming in regular classes. Besides, there are often problems with technical equipment and rarely proper system administrators. The teachers, who programmed in a formal setting, were concerned about the seriousness of the lessons and wondered if the students would actually learn the things they intended to. They also wondered how they would assess the students' results. Some concerns were expressed that it would be a pity to force a creative activity like programming into the framework of a regular school subject.

The advantages of a non-formal setting result from the disadvantages of the formal one. There is a fixed time frame available and there is no need to link the lessons to the curriculum. One could focus on fun and motivation of the children without the pressure of achieving predetermined learning goals. Additionally, one can control the size of the group and encourage only suitable or interested students to join the club. As a major downside of implementing programming in a non-formal environment, teachers point out that not all students are given the opportunity to participate.

Concerning our methodology—the exploratory interview–we can say that it was well suited for our purpose. We wanted to create a pleasant atmosphere for the teachers in which they could freely share their opinions and views with us. The rather open interview situation was suitable for this. However, we also think that it is difficult to create this atmosphere if you don't know each other at all. It was helpful that we knew the teachers beforehand. It was only possible in some cases to interview the teachers of the individual schools separately. When analyzing the interviews, however, we determined that the speech proportions in the group interviews were balanced and that the respective teachers also expressed very different opinions.

6.7 Conclusions and Future Directions

With the introduction of new curricula covering CS and computational thinking and the growing market of out-of-school coding activities for children, it is important to include the opinions of experts in the field—primary school teachers.

In our interviews, teachers mentioned some concerns and challenges of implementing programming in a formal setting, but these were mostly of a more practical nature and related to the concrete implementation in individual schools. When it came to whether they found it useful for the students, almost all of them agreed that all students should have the opportunity to learn programming. The fact that programming is not included in the Bavarian primary school curriculum is a (mostly time-related) problem for many teachers and should not be underestimated.

When the project is finished, we will make recommendations to the Bavarian Ministry of Education on how programming could be implemented in primary schools and where teachers would draw the line between formal and informal education. In our future work, we will try to revise the course concept according to the teachers' remarks. For example, more programming units could be developed that relate directly to existing parts of the curriculum.

Acknowledgments We would like to express our special thanks to all teachers involved in the project for their openness, curiosity, and commitment.

References

Bell, T., & Duncan, C. (2018). Teaching computing in primary schools. In S. Sentance, E. Barendsen, & C. Schulte (Eds.), *Computer science education*. Bloomsbury Academic.

Bergner, N., Köster, H., Magenheim, J., Müller, K., Romeike, R., Schroeder, U., et al. (2017). Zieldimensionen für frühe informatische Bildung im Kindergarten und in der Grundschule. In I. Diethelm (Ed.), *Informatische Bildung zum Verstehen und Gestalten der digitalen Welt* (pp. 15–24). Gesellschaft für Informatik.

Berry, M. (2015). *QuickStart primary handbook*. BCS.

Best, A. (2019). Bild der Informatik von Grundschullehrpersonen: Ergebnisse eines mehrjährigen Projekts zu informatikbezogenen Vorstellungen. In A. Pasternak (Ed.), *Informatik für alle* (pp. 59–68).

Best, A., Borowski, C., Büttner, K., Freudenberg, R., Fricke, M., Haselmeier, K., et al. (2019). Kompetenzen für informatische Bildung im Primarbereich. *LOG IN, 38*(1), 1–36.

Black, J., Brodie, J., Curzon, P., Myketiak, C., McOwan, P. W., & Meagher, L. R. (2013). Making computing interesting to school students: Teachers' perspectives. In J. S. Downie (Ed.), *Proceedings of the 13th ACM/IEEE-CS Joint Conference on Digital Libraries*, pp. 255–260. ACM.

Brown, N. C. C., Sentance, S., Crick, T., & Humphreys, S. (2013). Restart: The resurgence of computer science in UK Schools. *ACM Transactions on Computing Education, 1*(1).

Corbin, J. M., & Strauss, A. (1990). Grounded theory research: Procedures, canons, and evaluative criteria. *Qualitative Sociology, 13*(1), 3–21. https://doi.org/10.1007/BF00988593

Deutschschweizer Erziehungsdirektorenkonferenz. (2016). Medien und Informatik. In D-EDK (Ed.), *Lehrplan 21*. https://v-ef.lehrplan.ch/lehrplan_printout.php?e=1&k=1&fb_id=10

Diethelm, I., & Schaumburg, M. (2016). IT2School—development of teaching materials for CS through design thinking. In A. Brodnik & F. Tort (Eds.), *Informatics in schools: Improvement of informatics knowledge and perception* (Vol. 9973, pp. 193–198). Springer. https://doi.org/10.1007/978-3-319-46747-4

Duncan, C., Bell, T., & Atlas, J. (2017). What do the teachers think? Introducing computational thinking in the primary school curriculum. In D. Teague & R. Mason (Eds.), *Proceedings of the Nineteenth Australasian Computing Education Conference (ACE 2017)* (pp. 65–74). The Association for Computing Machinery. https://doi.org/10.1145/3013499.3013506

Duncan, C., Bell, T., & Tanimoto, S. (2014). Should your 8-year-old learn coding? In *Proceedings of the 9th Workshop in Primary and Secondary Computing Education*, pp. 60–69. https://doi.org/10.1145/2670757.2670774

Engeser, S., Limbert, N., & Kehr, H. (2008). *Studienwahl Informatik: Abschlussbericht zur Untersuchung*.

Falkner, K., Vivian, R., & Falkner, N. (2014). The Australian digital technologies curriculum: Challenge and opportunity. In J. Whalley & D. D'Souza (Eds.), *Proceedings of the Sixteenth Australasian Computing Education Conference*. ACM.

Friend, M., Matthews, M., Winter, V., Love, B., Moisset, D., & Goodwin, I. (2018). Bricklayer: Elementary students learn math through programming and art. In T. Barnes, D. Garcia, E. K. Hawthorne, & M. A. Pérez-Quiñones (Eds.), *Proceedings of the 49th ACM Technical Symposium on Computer Science Education—SIGCSE'18* (pp. 628–633). ACM Press. https://doi.org/10.1145/3159450.3159515

Funke, A., Berges, M., & Hubwieser, P. (2016a). Different perceptions of computer science. In S. Iyer & N. Thota (Eds.), *2016 International Conference on Learning and Teaching in Computing and Engineering (LaTICE)* (pp. 14–18). IEEE. https://doi.org/10.1109/LaTiCE.2016.1

Funke, A., Geldreich, K., & Hubwieser, P. (2016b). Primary school teachers' opinions about early computer science education. In *Proceedings of the 16th Koli Calling International Conference on Computing Education Research—Koli Calling'16* (pp. 135–139). https://doi.org/10.1145/2999541.2999547

Gärtig-Daugs, A., Weitz, K., Wolking, M., & Schmid, U. (2016). Computer science experimenter's kit for use in preschool and primary school. In J. Vahrenhold & E. Barendsen (Eds.), *Proceedings of the 11th Workshop in Primary and Secondary Computing Education* (pp. 66–71). ACM. https:// doi.org/10.1145/2978249.2978258

Geldreich, K., Funke, A., & Hubwieser, P. (2016). A programming circus for primary schools. In A. Brodnik & F. Tort (Eds.), *Informatics in schools: Improvement of informatics knowledge and perception* (pp. 46–47). Springer.

Geldreich, K., Simon, A., & Hubwieser, P. (2019). A design-based research approach for introducing algorithmics and programming to Bavarian primary schools. *MedienPädagogik: Zeitschrift Für Theorie Und Praxis Der Medienbildung, 33*(Medienpädagogik und Didaktik der Informatik), 53–75.

Geldreich, K., Talbot, M., & Hubwieser, P. (2018). Off to new shores: Preparing primary school teachers for teaching algorithmics and programming. In *Proceedings of the 13th Workshop in Primary and Secondary Computing Education on—WiPSCE'18*, pp. 1–6. https://doi.org/10.1145/ 3265757.3265783

Goecke, L., & Stiller, J. (2018). Informatische Phänomene und Sachunterricht. Beispiele für vielper-spektivischen Umgang mit einem Einplatinencomputer. In M. Thomas & M. Weigend (Eds.), *Informatik und Medien: 8. Münsteraner Workshop zur Schulinformatik*. Books on Demand.

Honer, A. (2011). Das explorative Interview: Zur Rekonstruktion der Relevanzen von Expertinnen und anderen Leuten. In A. Honer & R. Hitzler (Eds.), *Kleine Leiblichkeiten* (pp. 41–58). VS Verlag für Sozialwissenschaften/Springer Fachmedien Wiesbaden GmbH Wiesbaden.

Kultusministerkonferenz (Ed.). (2017). *Strategie der Kultusministerkonferenz „Bildung in der digitalen Welt". Beschluss der Kultusministerkonferenz vom 08.12.2016 in der Fassung vom 07.12.2017*. KMK. https://www.kmk.org/fileadmin/Dateien/veroeffentlichungen_beschluesse/ 2018/Strategie_Bildung_in_der_digitalen_Welt_idF._vom_07.12.2017.pdf

Kwon, S., & Schroderus, K. (2017). *Coding in schools: Comparing integration of programming into basic education curricula of Finland and South Korea*. Finnish Society on Media Education.

Lunenburg, F. C. (2010). Extracurricular activities. *Schooling, 1*(1).

Magenheim, J., Schulte, C., Schroeder, U., Humbert, L., Müller, K., Bergner, N., & Fricke, M. (2018). Das Projekt Informatik an Grundschulen. *LOG IN Informatische Bildung Und Computer in Der Schule, 189*(190), 57–66.

Maloney, J., Resnick, M., Rusk, N., Silverman, B., & Eastmond, E. (2010). The scratch programming language and environment. *ACM Transactions on Computing Education, 10*(4), 1–15. https://doi. org/10.1145/1868358.1868363

Moorman, P., & Johnson, E. (2003). Still a stranger here: Attitudes among secondary school students towards computer science. *ACM SIGCSE Bulletin, 35*(3), 193. https://doi.org/10.1145/961290. 961564

Sentance, S., Waite, J., Yeomans, L., & MacLeod, E. (2017). Teaching with physical computing devices. In *Proceedings of the 2017 ACM Conference on Innovation and Technology in Computer Science Education*, pp. 87–96. https://doi.org/10.1145/3137065.3137083

Topi, H. (2015). Gender imbalance in computing. *ACM Inroads, 6*(4), 22–23. https://doi.org/10. 1145/2822904

Ullrich, P. (2006). *Das explorative ExpertInneninterview*. Technische Universität Berlin. https:// doi.org/10.14279/DEPOSITONCE-4745

Walsh, I., Holton, J. A., Bailyn, L., Fernandez, W., Levina, N., & Glaser, B. (2015). What grounded theory is…A critically reflective conversation among scholars. *Organizational Research Methods, 18*(4), 581–599. https://doi.org/10.1177/1094428114565028

Webb, M., Davis, N., Bell, T., Katz, Y. J., Reynolds, N., Chambers, D. P., et al. (2017). Computer science in K-12 school curricula of the 21st century: Why, what and when? *Education and Information Technologies, 22*(2), 445–468. https://doi.org/10.1007/s10639-016-9493-x

Weng, X., & Wong, G. K. W. (2017). Integrating computational thinking into English dialogue learning through graphical programming tool. In *2017 IEEE 6th International Conference on*

Teaching, Assessment, and Learning for Engineering (TALE), pp. 320–325. https://doi.org/10. 1109/TALE.2017.8252356

Wing, J. M. (2006). Computational thinking. *Communications of the ACM, 49*(3), 33. https://doi. org/10.1145/1118178.1118215

Yadav, A., Gretter, S., Hambrusch, S., & Sands, P. (2016). Expanding computer science education in schools: Understanding teacher experiences and challenges. *Computer Science Education, 26*(4), 235–254. https://doi.org/10.1080/08993408.2016.1257418

Yadav, A., Mayfield, C., Zhou, N., Hambrusch, S., & Korb, J. T. (2014). Computational thinking in elementary and secondary teacher education. *ACM Transactions on Computing Education, 14*(1), 1–16. https://doi.org/10.1145/2576872

Katharina Geldreich studied primary school education at the University of Education Freiburg (2008–2013). Subsequently, she completed the cross-university masters program Media in Education (2013–2015) at the University of Education Freiburg and the University of Applied Sciences Offenburg. She has been a researcher at the Chair of Computer Science Education since 2016. Her work initially encompassed the development and implementation of new CS course concepts for primary and lower secondary school. Since 2018, she has been working on the research project AlgoKids, in which a total of 40 teachers from all over Bavaria are trained in algorithmics and programming and accompanied in their first attempts at teaching.

Peter Hubwieser Peter Hubwieser was teaching Mathematics, Physics, and Computer Science at Bavarian Gymnasiums for 15 years. In 1995, he received his Dr. rer. nat. in Theoretical Physics at the Ludwig-Maximilians-Universität München. From 1994 to 2002, he has been delegated to the Faculty of Informatics of the Technical University of Munich on behalf of the Bavarian Ministry of Education, in charge of the implementation of new courses of studies for teacher education in Informatics and a compulsory subject of Computer Science at Bavarian secondary schools. Since June 2002, he is working as an associate professor at the Faculty of Informatics of the Technical University of Munich. From 2002 to 2015, he was additionally visiting Professor at the Alpen-Adria-University of Klagenfurt, 2007 at the University of Salzburg, and 2008 at the University of Salzburg. In 2006, he received the Bavarian State Award for Education and Culture and 2011 the second-place award in the nationwide Contest for Subject Matter Didactics of the Polytechnische Gesellschaft Frankfurt. Since 2009, he holds the position of the Information Officer of the TUM School of Education. In 2015, he took over the position of the scientific director of the Schülerforschungszentrum Berchtesgaden additionally to his professorship.

Chapter 7
Games for Artificial Intelligence and Machine Learning Education: Review and Perspectives

Michail Giannakos, Iro Voulgari, Sofia Papavlasopoulou, Zacharoula Papamitsiou, and Georgios Yannakakis

Abstract Digital games have gained significance as a new paradigm in education. Digital games are accessible and affordable to anyone and provide opportunities for at-scale teaching and learning. In recent years, there has been increasing interest in digital games to support computational thinking and programming in pre-college (K–12) schools. Artificial Intelligence (AI) and Machine Learning (ML) are a rapidly developing field, attracting an increasing number of learners in the past few years. Although the confluence of digital games and AI/ML is an important and challenging field for teaching and learning researchers, a literature review has not yet been conducted in this area. The purpose of this work is to present a review of recent research into games to support AI and ML education. After a thorough search, relevant papers and games were selected and included in our qualitative content analysis. Based on this review, we present an overview of the relevant research papers and games, as well as showcased how different games provide a unique opportunity to teach a number of different concepts and topics in AI and ML.

Keywords Educational games · AI education · Machine learning education · Literature review

7.1 Introduction

During the last few years, digital games have become increasingly popular in Computer Science (CS) and Information Technology (IT) education (Harteveld et al. 2014; Kordaki and Gousiou 2016). Digital games have been a popular approach to

M. Giannakos (✉) · S. Papavlasopoulou · Z. Papamitsiou
Norwegian University of Science and Technology (NTNU), Trondheim, Norway
e-mail: michailg@ntnu.no

I. Voulgari · G. Yannakakis
Institute of Digital Games, University of Malta, Msida, Malta

I. Voulgari
National and Kapodistrian University of Athens, Athens, Greece

© Springer Nature Singapore Pte Ltd. 2020
M. Giannakos (ed.), *Non-Formal and Informal Science Learning
in the ICT Era*, Lecture Notes in Educational Technology,
https://doi.org/10.1007/978-981-15-6747-6_7

several endeavors to enhance CS education. At K–12 schools, some programs engage students in playing games that include tasks and problems that must be solved to progress (Vahldick et al. 2014), or encourage students to develop games using visual and block-based programming environments such as Alice (Cooper 2010) or Scratch (Resnick et al. 2009). However, most of the game-based learning efforts focus on the craft and practice of programming, rather than higher level CS concepts (Garneli et al. 2015).

In general, students are positive about game-related projects or game-based learning in the course curriculum or informal learning (Vahldick et al. 2014; Wallace et al. 2010). Moreover, such approaches have a positive impact on students' learning and motivation (Papastergiou 2009). Leutenegger (2006) demonstrates how students regularly exceed project requirements in his game-based course, while the norm is that students just meet the specified requirements. Thus, in the CS education literature, the game-based approach seems to provide inherent benefits and justifies the intense utilization of such practices in the CS education discipline (Vihavainen et al. 2014).

Our motivation for this work lies in the natural connection between games and artificial intelligence (AI) methods. In particular, the emphasis on the introduction of game elements in CS education has focused on dedicated game design and development courses, as well as on introductory courses (e.g., CS0, CS1, CS2), with great success (Vahldick et al. 2014; Wallace et al. 2010). In addition, games and puzzles have a long history as interesting problem domains for AI research (Wallace et al. 2010). Moreover, games have long been seen as the perfect test bed for AI methods (Yannakakis and Togelius 2018); therefore, the confluence of game elements and the AI domain is meaningful and helpful for students to develop an interest and competence in the increasingly important field of AI.

Keeping the aforementioned benefits and challenges in mind, this chapter centers on a literature review with the goal to present an overview of recent research into games to support AI and machine learning (ML) education. AI is expected to play an even more pervasive and critical role in education. In 2018, UNICEF launched the "Generation AI" initiative (https://www.unicef.org/innovation/GenerationAI), aiming to address and discuss the challenges and opportunities emerging in the face of AI advances while limiting the risks and safeguarding the rights of children. A recent working paper by UNESCO (Division for Policies and Lifelong Learning Systems, UNESCO's Education 2019) discusses the design of learning environments and learning management systems integrating AI, and the potential and challenges of AI for all education stakeholders such as students, teachers, administrators, and policymakers. Social and ethical concerns are raised, and the importance of the involvement of all stakeholders at the early stages of design rather than as mere beneficiaries or users is proposed. This chapter is situated in this field, the provision of tools for empowering students and educators to understand and become active participants in the design of AI and ML systems, and presents a general review of what game elements have been used to support AI and ML pre-college education and how. Although the confluence of game elements and AI/ML pre-college education is a relatively young area, enough work has already been done to conduct a review and provide insights.

7.2 Related Work

AI and ML are a rapidly developing field, attracting an increasing number of researchers and learners in the past few years. In response to this need, efforts in the USA, China, and many other countries are being developed to support AI education in K–12 schools (Touretzky et al. 2019a). In addition, during the last few years, new curricula and online resources have been developed, focusing on pre-college students and professional development for K–12 teachers to learn the basics of AI (Touretzky et al. 2019a). In 2018, the Association for the Advancement of Artificial Intelligence (AAAI) and the Computer Science Teachers Association (CSTA) announced a joint initiative to develop national guidelines for supporting AI education among K–12 students. Moreover, initiatives such as the AI for K–12 working group (AI4K12) and AI4All (https://ai-4-all.org) were established to define what students should know and be able to do with AI, as well as to develop national guidelines and collect resources (e.g., videos, demos, software, and activity descriptions) for AI education in the USA (Touretzky et al. 2019b).

During the last few years, several software and hardware tools have been created to allow young students to engage with AI and ML. For example, Cognimates (https://cognimates.me) offers a set of Scratch extensions to provide access to speech generation, speech recognition, text categorization, object recognition, and robot control Application Programming Interfaces (APIs). Kahn and Winters (2017) developed a similar set of extensions called eCraft2Learn (https://ecraft2learn.github.io/ai). The ML for Kids portal (https://machinelearningforkids.co.uk) provides online demos for students to train classifiers using web apps or Scratch extensions. Google has also developed several software tools to support students in engaging with AI concepts. For example, it has developed the concept of an "online AI experiment" (https://experiments.withgoogle.com/collection/ai), which allows young students to train visual classifiers (i.e., Teachable Machine) or see how a neural network tries to guess what they are drawing (i.e., QuickDraw). Another example is Google's "AI and You" kits that offer affordable Raspberry Pi Zero-based image and speech recognition (using a neural network classifier). Another example that allows K–12 students to explore neural networks and back-propagation learning via an interactive graphical tool is TensorFlow Playground (https://playground.tensorflow.org). Therefore, during the last few years, we have seen several initiatives resulting in software and hardware tools to support K–12 students in engaging with AI and ML.

Despite the rapid development of AI/ML education, novices find it hard and obscure to learn the fundamentals, such as game theory, machine learning, decision trees, and so on. During 2014, Educational Advances in Artificial Intelligence (EAAI) conference, 68% of the participants indicated games and puzzles as a topic they teach in their AI courses (Wollowski et al. 2016). In addition, Eaton et al. (2018) indicate that the introduction of agent-based models through games and puzzles allows instructors to introduce concepts for later exploration such as search, string-replacement iteration, planning, ML, and so on. It can, therefore, be agreed that games have long been seen as an ideal test-bed for understanding AI methods (Yannakakis and Togelius 2018).

Games (and game-based curricula) provide a widespread medium to support the teaching and learning of CS and IT (Vihavainen et al. 2014). Games have been used to improve several aspects in CS and IT education, for example, the lack of diversity in science, technology, engineering, and mathematics (STEM) fields, including CS, at both university and K–12 levels (Horn et al. 2016). Games have also been used as a means to enhance student engagement and motivation (Wallace et al. 2010). Another example comes from Clarke and Noriega (2003), who developed a war strategy game with hooks for the addition of AI modules. Their results indicate that students find AI much more interesting and accessible with examples and projects based on this game. Besides efforts to develop novel game-based curricula to enhance teaching and learning CS in the context of formal education, many efforts to reach younger students are made in informal contexts such as classroom visits, summer camps, and after-school programs (Vahldick et al. 2014).

In addition to several beneficial qualities of games, such as engagement, competition, and collaboration, they lead to greater student interest (Horn et al. 2016; Papastergiou 2009). AI and ML, as content, certainly can benefit from games (e.g., a game engine that can be easily used or modified; Hartness 2004). Moreover, Cook and Holder (2001) used a simple game to teach students about the need for internal representations of the world, natural language processing, look-ahead search, plan generation, and ML, demonstrating the power of games to support AI/ML education. Their students managed to significantly improve and modify the game to handle different problems.

In our view, ML and AI education can significantly benefit by introducing to students state-of-the-art algorithms, concepts, and methods related to game playing, for example, game tree-based search, reinforcement learning, and neural networks. Such an approach has tremendous potential to successfully introduce those concepts to pre-college students. Looking in the literature, we can find several studies that exploit the intersection between games and AI/ML to form the basis for a dedicated course or a module of a course (Zhou et al. 2017, 2018; Li et al. 2019; Konen 2019). In this work, we provide a general review of games and software tools that can be used to support AI and ML pre-college education. This collection will provide a springboard for other scholars and practitioners to put into practice, experiment, compare, and adapt the games and software listed to meet the needs of their students.

7.3   Methods

To the best of our knowledge, no previous work has aimed to produce an overview of games and software tools that can be used to support AI and ML pre-college education. Thus, the aim of this chapter is to collect and summarize the various games and software tools. The comprehensive review provided in this chapter could help and guide different stakeholders to explore and put into practice the games that meet their needs.

The selection phase determines the overall validity of the work, and thus it is important to define specific criteria. Games were eligible for inclusion if they focused on AI/ML education. To find those games, we searched various libraries and search engines (Google) as well as scientific publications (e.g., Google Search, ACM Digital Library, IEEE Xplore, Science Direct, Google Scholar).

The search string used during the search covers two main terms for content ("AI Education," "ML Education," "CS Education") and the medium ("Game-Based Learning," "Games for Learning"). The combination resulted in six different search strings. Due to the high number of irrelevant papers (i.e., false positives) returned using the search string "CS Education," the authors decided to narrow the search by combining it with the term "AI" or "ML."

7.4   Findings

Finally, after implementing the aforementioned search strategy, we reviewed the outcomes of the search and identified 17 games/projects. Then, we reviewed those games and projects and summarized their essential elements and focus (Table 7.1). These summaries allowed us to consolidate the essence and the main focus of the games/projects and their connection with AI/ML concepts.

In Table 7.1 we summarize games and platforms for supporting AI/ML pre-college education. Many of them focus on the wider area of CS education, with applications in AI/ML education as well. In particular, we identified only a small number of games, applications, and platforms specifically aimed at explicitly supporting AI and ML education for children and young people. Coding seems to be the main goal of most of the existing environments. However, environments aiming to enhance AI and ML pre-college education mainly address concepts such as training a model for image, text, or audio recognition (e.g., Machine Learning for Kids and AI Machine Learning Education Tools) and programming logic (e.g., Minecraft Hour of Code: AI for Good), while games not aimed at formal children's education address more abstract, ethical, and social implications (e.g., The Moral Machine, Universal Paperclips).

When it comes to the age those environments are focusing on, we found that they address the whole range from kindergarten to high school (K–12), with some of them addressing even younger ages (appropriate for 4 years old). In addition, there are environments supporting parents and teachers in teaching AI and ML to children

Table 7.1 Summary of digital games for Artificial Intelligence (AI) and Machine Learning (ML) education

Name of the game	Short description	Reference
Bug Brain	Bug Brain is a game where children can experiment with the neurons and nodes that make up a brain. They build a brain for a ladybug to help it feed and survive. Not specifically aimed at learning AI and ML, Bug Brain features rendered graphics, challenging puzzles, and the opportunity to learn about neural networks (free)	https://www.biologic.com.au/bugbrain
Human Resource Machine	Human Resource Machine is a puzzle game. Players are required to solve problems through programming. Concepts relevant to AI such as automation and optimization are introduced. At each level, players have to automate work by programming the employees of an office environment (purchase required)	https://tomorrowcorporation.com/humanresourcemachine
7 Billion Humans	Following up on the "Human Resource Machine" game and developed by the same studio, players are required to solve puzzles by programming multiple agents (workers). Concepts such as parallel computing, debugging, and optimization are explored (purchase required)	https://tomorrowcorporation.com/7billionhumans

(continued)

Table 7.1 (continued)

Name of the game	Short description	Reference
Machine Learning for Kids	Machine Learning for Kids introduces ML by providing hands-on experiences for training ML systems and building things with them. It provides an easy-to-use guided environment for training ML models to recognize text, numbers, images, or sounds. Machine Learning for Kids adds models to educational coding platforms Scratch and App Inventor, and helps children to create projects and build games with the ML models they train (free)	https://machinelearningforkids.co.uk
AI Machine Learning Education Tools	The platform (currently in beta) offers tools for teaching students the basic concepts of ML. It incorporates a number of Scratch extensions as a coding medium for children, such as a chatbot extension, home automation, image recognition, classification, and teaching the computer how to play the Flappy Bird game, accompanied by lesson plans and materials for educators (free registration required)	https://www.ai4children.org
While True: Learn()	This game aims to familiarize players with the concepts and processes of ML. Players take up the role of an ML specialist who uses visual programming to complete clients' projects. It includes elements such as neural networks, actual ML techniques, and ML-related problems such as self-driving cars (purchase required)	https://luden.io/wtl

(continued)

Table 7.1 (continued)

Name of the game	Short description	Reference
ViPER	ViPER aims to teach concepts in ML to middle-school students. By programming a robot to solve pathfinding problems, players learn how machines learn, and engage with concepts such as algorithms, the testing, and training phase, and identifying patterns in the data (for assessment of game design issues, see also Parker and Becker 2014; purchase required)	https://wonderville.org/asset/ViPER
Minecraft Hour of Code: AI for Good	The game integrates a coding interface with Minecraft. By programming a robot to predict forest fires, players are introduced to basic coding concepts and learn about AI and its potential for protecting the environment. A lesson plan and supporting material for educators are also provided (free)	https://education.minecraft.net/hour-of-code
The Moral Machine	The players are asked to choose the lesser evil when facing an impending car crash. This platform is mainly situated in the field of ethical decision-making. It aims to address the diversity of human perspectives in the face of a moral dilemma and the implications of machine intelligence designed to make similar moral decisions (e.g., self-driving cars; free)	https://moralmachine.mit.edu

(continued)

Table 7.1 (continued)

Name of the game	Short description	Reference
PopBots (Preschool-Oriented Programming Platform)	The applications included in this platform aim to familiarize young children with the main concepts and processes of AI, such as programming, classification, training, and testing datasets, through simple activities such as training models to recognize healthy and unhealthy food or different types of music. It also includes supporting material (e.g., lesson plans) for teachers (for more details and related study, see also Williams et al. 2019a, b)	https://www.media.mit.edu/projects/pop-kit/ove rview
Universal Paperclips	Not quite a game for teaching AI concepts, but rather for triggering discussion on the role and potential of AI in society. Based on the philosophical thought experiment "paperclip maximizer" about AI design and machine ethics, this is a clicker game where the player takes up the role of an AI machine making paperclips. After several upgrades such as the possibility to "interpret and understand human language" or buy "autonomous aerial brand ambassadors," the game ends when 100% of the universe is explored and all matter is turned to paperclips (web version free)	https://www.decisionproblem.com/paperclips
Gladiabots: AI combat arena	A game not specifically aiming to teach AI to students, but as the designers describe, the players have to assemble a "perfect team of robots and set their AI strategy with the simple to use but satisfyingly deep visual AI editor." Players are introduced to the logic and structure of AI programming (purchase required)	https://gladiabots.com

(continued)

Table 7.1 (continued)

Name of the game	Short description	Reference
Tynker: coding for kids	Platform including applications and games, separated by age groups, for children as young as 5. Children can create games through block programming, and share their artifacts (subscription required)	https://www.tynker.com
Scratch Jr.	The younger version of Scratch, aiming at children ages 5–7. Through a simple, visual, drag-and-drop interface, children create code, program, and share their own projects with the community (free, available for Android and iOS devices)	https://www.scratchjr.org
Code.org	The platform, aimed at children and educators, includes applications, games, and courses for learning coding, creating new projects, and sharing them with the community, as well as a curriculum and lesson plans for CS education. Code.org also organizes the annual Hour of Code campaign engaging students in coding around the world (free)	https://code.org
LightBot	LightBot is a puzzle game based on coding, aiming to teach programming logic to children as young as 4 and above. It is translated into multiple languages. Web (free), iOS and Android device (purchase required) versions	
Codespark Academy	A platform aiming at children aged 5–9 that includes games, puzzles, and applications for learning coding and creating new games, as well as resources for parents and educators (subscription required)	https://codespark.com

(e.g., lesson plans such as Minecraft Hour of Code: AI for Good). Although most of the materials have been implemented primarily in the English language, we also see environments and materials supporting multiple languages (a good example is Code.org).

Looking at the types of platforms utilized from the environments identified, we noticed that there is a wide variety, such as proper applications, applications running on the web, as well as applications that are developed for mobile devices such as mobile phones and tablets. As with any applications, those requiring installation (e.g., While True: Learn(), Human Resource Machine) are more robust and do not necessarily rely on an internet connection, compared to those that run online and do not require installation (e.g., The Moral Machine). Another important dimension of AI/ML learning environments is the cost. Looking the identified environments, we note that many of them are free, or have a free version (e.g., The Moral Machine, Code.org); however, there is also a reasonable number of games where a purchase or a paid subscription is required (e.g., Codespark Academy, Gladiabots: AI combat arena); in most cases, the teacher/parent can have a free trial with the game.

In order for pre-college students, instructors, and parents to understand the fundamental ideas of AI and ML, they also need to be able to engage with them practically. Most of the identified games have been developed during the last few years and schools and teachers have just started to adopt them. In the near future, we expect further development of the available environments, but also more environments to be accessible. Moreover, besides games, we have also seen an increasing number of daily products and tools that demonstrate AI's capabilities (Google Assistant, Apple's Siri, Microsoft's Cortana), and there are a number of home appliances with similar functionality (Google Home, Amazon Echo, Apple HomePod). Most of them are used by young children and will help them to familiarize themselves with AI technologies. Going a step further, a variety of new software and hardware tools are providing AI components to young programmers who can incorporate them into their own creations.

7.5 Discussion

Despite their increasing role in everyday life and society, AI and ML are not being fully explored in schools. Opportunities for teaching relevant skills and competences through novel approaches have the capacity to revolutionize the contemporary teaching of computational and algorithmic thinking and CS overall. Skills and competencies relevant to AI and ML, such as abstract thinking, problem-solving, and management of data and information, will empower students to adopt a more critical and inquisitive approach toward existing systems (e.g., recognizing bias, disinformation, biased search rankings, filter bubbles) and to participate in the design of new ones (Turchi et al. 2019).

Although games that support AI and ML seem to be in their infancy, in this literature review, we identified a good number of games and applications, for various ages, school levels, and learner expertise, aiming to teach AI and ML concepts to young children, by providing either guided environments for practice or more open-ended environments where children can create their own projects and creatively express their ideas. Although the number of environments specifically aiming to teach AI, ML, and related concepts to young children is still limited, it is steadily increasing, following the general interest (UNESCO's Education 2019). For instance, games such as PopBots, Minecraft Hour of Code: AI for Good, and While True: Learn(), and environments incorporating game elements such as the Teachable Machine, AI Machine Learning Education Tools, and Machine Learning for Kids, are indicative examples that have a particular focus on AI and ML concepts.

Both guided and open-ended environments have been identified in the literature. Both types can be used to support different learning designs and to scaffold AI and ML concepts. For instance, guided environments can help students by directing them to master concrete concepts, practices, and processes, while open-ended environments empower students to utilize and further their understanding of AI and ML concepts by deeply engaging in active learning and even by constructing artifacts.

Supporting material for students and educators is an extremely useful resource that can enhance the attainment of their learning objectives. AI and ML are still a new topic in pre-college education, therefore students and teachers require more than an educational game to approach, understand, and be able to discuss the relevant concepts. Learning about the subject matter by only playing a game may be challenging for both the student and the teacher, as well as insufficient for deep conceptual understanding, and might, therefore, lead students to develop misconceptions (Muehrer et al. 2012; Parker and Becker 2014). Environments such as Minecraft Hour of Code: AI for Good and AI Machine Learning Education Tools provide good practices by offering lesson plans, additional activities, and other resources for teachers. In this framework, Camilleri et al. (2019) recently published a practical guide, financed by the Ministry for Education of Malta, with lesson plans and resources for teachers aiming to teach AI to young people. It is important for both the student and the teacher to have proper learning designs and materials around these games that support AI and ML holistically.

Easy access, price, and technical requirements constitute further critical factors for the effectiveness, adoption, and impact of these learning environments. Not all pre-college education schools and families have the budget or the technological infrastructure and competence to access sophisticated games or platforms (Marklund and Taylor 2016). The effectiveness of a game-based curriculum in schools relies upon multiple context-related factors, such as the game literacy of students, the technological skills of teachers, class schedule restrictions, the computers available and their specifications, and the available bandwidth. Games with low technical demands and requiring fewer technical skills, such as Minecraft Hour of Code: AI for Good and Code.org, which require no installation, student accounts, or cutting-edge technology computers, seem more appropriate for formal school settings.

This preliminary work should not be seen as a systematic review, but as an early effort to provide a general overview and inspire instructors and future researchers.

Although we tried to identify most of the relevant games and projects, we recognize that different search strategies and selections (e.g., databases, query) might bring additional useful results. In addition, the selection of projects and games might also pose another possible limitation. However, the focus of the selected games and projects was clearly on AI and ML education; the summary was undertaken by two researchers and included the main qualities. Many of the reported games have not been extensively used and evaluated (as is the case in games reported in scientific publications), leading to some missing information about their effectiveness and acceptance by students. This is mainly based on the fact that AI and ML in K–12 schools are a relatively young field of research, and we expect to see more empirical studies and projects addressing these issues in the near future.

7.6 Conclusions

General game playing is an exciting topic, still young but on the verge of maturing, which touches upon a broad range of aspects of AI and ML. In this chapter, we created a general overview of games for pre-college AI/ML education, in an attempt to show its many facets and highlight the fact that it provides a rich source of interesting and challenging qualities for pre-college students and instructors who want to introduce their students to AI/ML concepts. We also showed how different games provide a unique opportunity to teach a number of different concepts and topics in AI and ML.

Although research on the use of games or other applications teaching AI and ML to children and young people is still very limited, early results show great potential for teaching even pre-school children basic AI and ML concepts, as well as for engaging them in conversations about the role and implications of technology and AI in our everyday lives (Druga et al. 2017; Williams et al. 2019a, b). Nevertheless, game design for engaging students and achieving an understanding of the concepts can be challenging, requiring appropriate metaphors and easy-to-understand interactions (Parker and Becker 2014). Children interact daily with applications and devices integrating AI (e.g., smart toys, smart home applications, video-sharing, and streaming platforms) with potential privacy, safety, and bias risks (McReynolds et al. 2017; Chu et al. 2019). Understanding the processes and factors involved in the design of such systems can help children to develop a more accurate mental model of their limitations and potential (of AI/ML).

Acknowledgements This work is supported by the "Learning science the fun and creative way: coding, making, and play as vehicles for informal science learning in the 21st century" Project, under the European Commission's Horizon 2020 SwafS-11-2017 Program (Project Number: 787476) and the "Learn to Machine Learn" (LearnML) project, under the Erasmus+ Strategic Partnership program (Project Number: 2019-1-MT01-KA201-051220).

References

Camilleri, V., Dingli, A., & Montebello, M. (2019). *AI in education: A practical guide for teachers and young people*. Department of AI: University of Malta.

Chu, G., Apthorpe, N., & Feamster, N. (2019). Security and privacy analyses of Internet of Things: Children's toys. *IEEE Internet of Things Journal, 6*(1), 978–985. https://doi.org/10.1109/JIOT.2018.2866423.

Clarke, D., & Noriega, L. (2003). Games design for the teaching of artificial intelligence. Interactive Convergence: Research in Multimedia, Aug. 7–9, Prague, Czech Republic.

Cook, D.J., & Holder, L.B. (2001). A client-server interactive tool for integrated artificial intelligence curriculum. In *Proceedings of the FLAIRS Special Track on AI Education*, pp. 206–210.

Cooper, S. (2010). The design of Alice. *ACM Transactions on Computing Education (TOCE), 10*(4), 15.

Division for Policies and Lifelong Learning Systems, UNESCO's Education. (2019). Artificial Intelligence in Education: Challenges and Opportunities for Sustainable Development (Working Paper No. 7; Working Papers on Education Policy, p. 48). UNESCO. https://unesdoc.unesco.org/ark:/48223/pf0000366994.

Druga, S., Williams, R., Breazeal, C., & Resnick, M. (2017). 'Hey Google is it OK if I eat you?' Initial explorations in child-agent interaction. *Proceedings of the 2017 Conference on Interaction Design and Children*, 595–600. https://doi.org/10.1145/3078072.3084330.

Eaton, E., Koenig, S., Schulz, C., Maurelli, F., Lee, J., Eckroth, J., et al. (2018). Blue sky ideas in artificial intelligence education from the EAAI 2017 new and future AI educator program. *AI Matters, 3*(4), 23–33

Garneli, V., Giannakos, M.N., & Chorianopoulos, K. (2015). Computing education in K-12 schools: A review of the literature. In *2015 IEEE Global Engineering Education Conference (EDUCON)*, pp. 543–551. IEEE.

Harteveld, C., Smith, G., Carmichael, G., Gee, E., & Stewart-Gardiner, C. (2014). A design-focused analysis of games teaching computer science. In *Proceedings of Games + Learning + Society*, 10, 1–8.

Hartness, K. (2004). Robocode: Using games to teach artificial intelligence. *Journal of Computing Sciences in Colleges, 19*(4), 287–291.

Horn, B., Clark, C., Strom, O., Chao, H., Stahl, A. J., Harteveld, C., & Smith, G. (2016). Design insights into the creation and evaluation of a computer science educational game. In *Proceedings of the 47th ACM Technical Symposium on Computing Science Education*, pp. 576–581. ACM.

Kahn, K., & Winters, N. (2017). Child-friendly programming interfaces to AI cloud services. In *European Conference on Technology Enhanced Learning*, pp. 566–570. Springer.

Konen, W. (2019). General board game playing for education and research in generic AI game learning. In *2019 IEEE Conference on Games (CoG)*, pp. 1–8. IEEE.

Kordaki, M., & Gousiou, A. (2016). Computer card games in computer science education: A 10-year review. *Journal of Educational Technology & Society, 19*(4), 11–21.

Leutenegger, S. T. (2006). A CS1 to CS2 bridge class using 2D game programming. *Journal of Computing Sciences in Colleges, 21*(5), 76–83.

Li, W., Zhou, H., Wang, C., Zhang, H., Hong, X., Zhou, Y., & Zhang, Q. (2019). Teaching AI algorithms with games including Mahjong and FightTheLandlord on the Botzone online platform. In *Proceedings of the ACM Conference on Global Computing Education*, pp. 129–135. ACM.

Marklund, B. B., & Taylor, A. S. A. (2016). Educational games in practice: The challenges involved in conducting a game-based curriculum. *Electronic Journal of e-Learning, 14*(2), 122–135.

McReynolds, E., Hubbard, S., Lau, T., Saraf, A., Cakmak, M., & Roesner, F. (2017). Toys that listen: A study of parents, children, and internet-connected toys. In *Proceedings of the 2017 CHI Conference on Human Factors in Computing Systems*, 5197–5207. https://doi.org/10.1145/3025453.3025735.

Muehrer, R., Jenson, J., Friedberg, J., & Husain, N. (2012). Challenges and opportunities: Using a science-based video game in secondary school settings. *Cultural Studies of Science Education, 7*(4), 783–805. https://doi.org/10.1007/s11422-012-9409-z.

Papastergiou, M. (2009). Digital game-based learning in high school computer science education: Impact on educational effectiveness and student motivation. *Computers & Education, 52*(1), 1–12.

Parker, J.R., & Becker, K. (2014). ViPER: Game that teaches machine learning concepts—a postmortem. In 2014 IEEE Games and Entertainment Media Conference (GEM).

Resnick, M., Maloney, J., Monroy-Hernández, A., Rusk, N., Eastmond, E., Brennan, K., et al. (2009). Scratch: Programming for all. *Communications of the ACM, 52*(11), 60–67.

Touretzky, D., Gardner-McCune, C., Breazeal, C., Martin, F., & Seehorn, D. (2019a). A year in K-12 AI education. *AI Magazine, 40*(4), 88–90. https://doi.org/10.1609/aimag.v40i4.5289.

Touretzky, D., Gardner-McCune, C., Martin, F., & Seehorn, D. (2019b). Envisioning AI for K-12: What should every child know about AI. In *Thirty-Third AAAI Conference on Artificial Intelligence (AAAI-19)*.

Turchi, T., Fogli, D., & Malizia, A. (2019). Fostering computational thinking through collaborative game-based learning. *Multimedia Tools and Applications.* https://doi.org/10.1007/s11042-019-7.

Vahldick, A., Mendes, A. J., & Marcelino, M. J. (2014). A review of games designed to improve intro-ductory computer programming competencies. In *2014 IEEE Frontiers in Education conference (FIE) Proceedings*, pp. 1–7. IEEE.

Vihavainen, A., Airaksinen, J., & Watson, C. (2014). A systematic review of approaches for teaching introductory programming and their influence on success. In *Proceedings of the Tenth Annual Conference on International Computing Education Research*, pp. 19–26. ACM.

Wallace, S. A., McCartney, R., & Russell, I. (2010). Games and machine learning: A powerful combination in an artificial intelligence course. *Computer Science Education, 20*(1), 17–36.

Williams, R., Park, H.W., & Breazeal, C. (2019a). A is for artificial intelligence: The impact of artificial intelligence activities on young children's perceptions of robots. In *Proceedings of the 2019 CHI Conference on Human Factors in Computing Systems—CHI'19*, pp. 1–11. https://doi.org/10.1145/3290605.3300677.

Williams, R., Park, H. W., Oh, L., & Breazeal, C. (2019b). PopBots: Designing an artificial intelli-gence curriculum for early childhood education. *Proceedings of the AAAI Conference on Artificial Intelligence, 33*(01), 9729–9736. https://doi.org/10.1609/aaai.v33i01.33019729.

Wollowski, M., Selkowitz, R., Brown, L.E., Goel, A., Luger, G., Marshall, J., et al. (2016). A survey of current practice and teaching of AI. In Thirtieth AAAI Conference on Artificial Intelligence.

Yannakakis, G. N., & Togelius, J. (2018). *Artificial intelligence and games* (Vol. 2). New York: Springer.

Zhou, H., Zhang, H., Zhou, Y., Wang, X., & Li, W. (2018). Botzone: An online multi-agent competi-tive platform for AI education. In *Proceedings of the 23rd Annual ACM Conference on Innovation and Technology in Computer Science Education*, pp. 33–38. ACM.

Zhou, H., Zhou, Y., Zhang, H., Huang, H., & Li, W. (2017). Botzone: A competitive and interactive platform for game AI education. In *Proceedings of the ACM Turing 50th Celebration Conference-China*, p. 6. ACM.

Michail Giannakos is a Professor of interaction design and learning technologies at the Depart-ment of Computer Science of NTNU, and Head of the Learner-Computer Interaction lab (https://lci.idi.ntnu.no/). His research focuses on the design and study of emerging technologies in online and hybrid education settings, and their connections to student and instructor experiences and practices. Giannakos has co-authored more than 150 manuscripts published in peer-reviewed jour-nals and conferences (including Computers & Education, Computers in Human Behavior, IEEE TLT, Behaviour & Information Technology, BJET, ACM TOCE, CSCL, Interact, C&C, IDC to mention few) and has served as an evaluator for the EC and the US-NSF. He has served/serves in various organization committees (e.g., general chair, associate chair), program committees as well as editor and guest editor on highly recognized journals (e.g., BJET, Computers in Human Behavior, IEEE TOE, IEEE TLT, ACM TOCE). He has worked at several research projects funded by diverse sources like the EC, Microsoft Research, The Research Council of Norway (RCN), US-NSF, German agency for international academic cooperation (DAAD) and Cheng Endowment; Giannakos is also a recipient of a Marie Curie/ERCIM fellowship, the Norwegian Young Research Talent award and he is one of the outstanding academic fellows of NTNU (2017–2021).

Iro Voulgari is a postdoctoral researcher at the Institute of Digital Games, University of Malta and teaching staff at the Department of Early Childhood Education, National and Kapodistrian University of Athens. She is teaching undergraduate and postgraduate courses on Digital Games and Virtual Worlds, and Learning Technologies. Her research focuses game based learning, game studies, and digital literacy. She has organised several workshops relevant to game based learning, Information and Communication Technologies in Education, and Digital Storytelling in local and international venues. She has worked on several Nationally and EU funded research projects on the design, implementation, and assessment of Learning Technologies in teaching and learning.

Sofia Papavlasopoulou is a post-doc researcher at the Department of Computer Science, Norwegian University of Science and Technology (NTNU). She holds a Ph.D. from NTNU in the area of Learning Technologies. Her interests center on the use of technological tools to support students' learning while enhancing their interest in Computing Education and coding, providing creative engagement. Her goal is to investigate the best ways to support an interactive, engaging approach to informal learning coding activities for young students and design meaningful learning experiences for them. Since 2015, she has co-authored more than twenty-five research articles in peer-reviewed journals and conferences. The results of her studies have been published in leading Journals like Computers in Human Behavior (Elsevier) and Entertainment Computing (Elsevier), International Journal of Child-Computer Interaction (Elsevier), and conferences like the ACM Conference on Interaction Design and Children (IDC).

Zacharoula Papamitsiou is a senior researcher at the Department of Computer Science (IDI), Norwegian University of Science and Technology (NTNU). She holds a Ph.D. degree from the University of Macedonia, Thessaloniki, Greece, in adapting and personalizing learning services for supporting learners' decision-making using Learning Analytics. Her research interest is on user modeling, quantified-self technologies, multimodal learning, human-computer interaction, and autonomous learning. She has worked on complex learning aspects (e.g., autonomous decision-making, engagement) using learning analytics for the quantification of learning behaviors and the interpretation of learning behavioral patterns. She has published articles in ranked international journals including CHB, BJET, IEEE TLT, ETR&D, and ET&S. She is a professional member of ACM, IEEE Technical Committee on Learning Technology, founding member of the Trondheim-ACM-W Chapter, and has served/serves in various organization committees (e.g., associate chair, workshop chair) and program committees. She is also recipient of the ERCIM fellowship (2018–2019).

Georgios Yannakakis is a Professor and Director of the Institute of Digital Games, University of Malta and a co-Founder of modl.ai. He is a leading expert of the game artificial intelligence research field with core theoretical contributions in machine learning, evolutionary computation, affective computing and player modelling, computational creativity and procedural content generation. He has published more than 250 papers and his work has been cited broadly. He has attracted funding from several EU and national research agencies and received multiple awards for published work in top-tier journals and conferences. His work has been featured in New Scientist, Science Magazine, The Guardian, Le Monde and other venues. He is regularly invited to give keynote talks in the most recognised conferences in his areas of research activity and has organised a few of the most respected conferences in the areas of game AI and game research. He has been an Associate Editor of the IEEE Transactions on Computational Intelligence and AI in Games and the IEEE Transactions on Affective Computing journals; he is currently an Associate Editor of the IEEE Transactions in Games. He is the co-author of the Artificial Intelligence and Games textbook.

Part IV
Learning Design and Experience

This part provides insight into different aspects of learning design, learner experience and engagement.

Chapter 8
Looking at the Design of Making-Based Coding Activities Through the Lens of the ADDIE Model

Sofia Papavlasopoulou and Michail Giannakos

Abstract *Making* has received growing interest in formal and informal learning environments. However, there is an acute need to investigate and get a deep understanding of the characteristics of making-based coding activities for children and how to appropriately design them. Over 3 years, we conducted empirical studies to investigate children's learning experience during making-based coding workshops, in which children used a block-based programming environment (i.e., Scratch) and collaboratively created a socially meaningful artifact (i.e., a game). This chapter aims to illustrate and discuss the learning design, using the ADDIE instructional model, and lessons learned based on a making-based coding workshop in Norway, named Kodeløypa.

Keywords Coding · Making · ADDIE model · Instructional design · Instructional model · Children

8.1 Introduction

Instructional design (ID) is a systematic process of designing the instruction of a learning event in an efficient manner. The ID process consists of phases that aim to investigate and determine learning objectives; develop learning materials, strategies, and assessment tools for evaluation; and accommodate an environment that encompasses successful learning outcomes (Morrison et al. 2019). Different ID models exist, with many of them based on the generic ADDIE model (**A**nalysis, **D**esign, **D**evelopment, **I**mplementation, and **E**valuation), an instructional model that describes a step-by-step process that can be used by instructional designers and practitioners who want to plan and create educational training and learning events. It

S. Papavlasopoulou (✉) · M. Giannakos (✉)
Norwegian University of Science and Technology (NTNU), Trondheim, Norway
e-mail: spapav@ntnu.no

M. Giannakos
e-mail: michailg@ntnu.no

© Springer Nature Singapore Pte Ltd. 2020　　　　　　　　　　　　　　　137
M. Giannakos (ed.), *Non-Formal and Informal Science Learning
in the ICT Era*, Lecture Notes in Educational Technology,
https://doi.org/10.1007/978-981-15-6747-6_8

presents a dynamic and flexible tool that can be adapted and used in many different contexts and has been widely applied in various educational projects (Morrison et al. 2019). The model was developed in the 1990s by Reiser and Mollenda and has five phases: Analysis, Design, Development, Implementation, and Evaluation. These phases describe specific actions and clear instructions that are simple and easy to adopt, but at the same time are quite generic. It is also possible to use the ADDIE model as a framework for the development of educational products (Alley and Jansak 2001) and to provide a systematic approach that can be integrated into learning strategies (Hall 1997; Pribadi 2009).

Using the ADDIE model for ID provides a basis to determine—depending on the course and the context that is applied each time—the learning objectives, develop the activities of a course, and evaluate the learners' progress and the effectiveness of the instruction. In the analysis phase, the starting point of the ADDIE model, specialists should investigate and have a clear view of what the learners already know, define the course's needs and characteristics, and develop instructional strategies. The next phase is design, which deals with the learning objectives, the content, the planning of the course, and the media selection. Drawing upon all the knowledge gained and the decisions made in the previous phases; in the development phase, course content and learning materials are created, assembling the resources that were created in the previous phase. Depending on the course, the ADDIE implementation phase may include management issues, but it basically aims to put into action the plan decided in the previous phases, evaluate its effectiveness, and ensure that everything performs as planned. Lastly, the evaluation phase represents a process that can happen at any of the stages of the ID process and aims to get feedback for improvement of the instruction and the materials and to confirm that the learning goals and objectives of the course are met. Overall, it is important that the process during all the phases is systematic and specific to achieve the course's goals.

The purpose of this chapter is to frame a making-based coding activity that takes place in an informal setting, using the ADDIE instructional model. Linking these activities with an existing model provides a systematic approach to design; this action can respond to a corresponding lack of improvement in learning practices and outcomes, and contribute to the design of meaningful learning experiences for specific needs and contexts. In addition, when instructors are in a mindset that allows them to think in such a way that they can structure their intuitive decisions, by interacting using a specific model and theory they can reflect, understand, and consequently make the design of the activity better.

8.2 The ADDIE Instructional Model and Its Application in Coding-Related Activities

The ADDIE model has been extensively used to meet the needs of learning events related to coding activities. It has also been used as a development process for materials and software tools related to learning. In their study, Novák et al. (2018) have used the ADDIE model for the design of educational materials, supporting the use of the Arduino platform, for teaching coding in high schools. The five phases of the model helped them to use a strategy for the development of the educational materials; through the analysis, they recognized the tasks that it is appropriate to include in the materials. Then, the learning materials were divided into lesson guides. Those had proposals with tasks that the teachers can do, including worksheets, where the focus is primarily on students, and, depending on the topic, each time they included relevant questions. The ADDIE model has also been applied for the development of multimedia instructional material for robotics education (Liu et al. 2008). Such materials are designed to engage students through an adventure story in the assembly of a robot and the coding of its operations to complete the mission of the story. Aiming to support university students and teachers with Object-Oriented Programming (OOP) learning, Oliveira and Bonacin (2018) suggested the design and implementation of OOP learning tasks with digital modeling and fabrication. The ID is based on the ADDIE model, offering a systematic process in this challenging project of using such technologies in formal educational settings.

The different settings in which the ADDIE model is applied are reflected also in its use for developing different kinds of multimedia. One example is an adventure game to support students' understanding of basic programming in vocational high school (Hidayanto et al. 2017). Based on the ADDIE model's five stages, the authors created and evaluated their game with students, measuring their learning based on their understanding of programming, and evaluating the software and visual communication. Similarly, Salahli et al. (2017), following the ADDIE model, developed a mobile application for the Scratch programming environment, supporting secondary school students to enhance their programming skills. In the analysis stage of the model, the authors not only analyzed the affordances of the Scratch programming language but also determined their target group of students. After the design and development phase, the students tested the mobile application in the implementation phase using pre–post skill tests. Based on the results, students from the experimental group who used the mobile application had a significant increase in their programming skills over those in the control group.

The ADDIE model has very often been modified in practice in compliance with the different learning settings that are applied. Wu (2014) proposed a seven-phase ID model based on ADDIE for educating game programmers. The goal is to create a model that is customized to the needs of stakeholders, curriculum developers, content designers, and others. In that case, the seven phases included "Definition" (providing a clear goal), planning and verification (to meet the industry's expectations), Design, Development, Implementation, and "Continuous Improvement" (continuous reevaluation and redesign of the instructional content to fulfill changing requirements).

The ADDIE model has been successfully associated with good quality design; definition of clear objectives; appropriately designed materials, media, and content; a well-arranged workload for teachers and students; and evaluation connected to the desired outcomes. Thus, supporting the design of informal educational settings with ADDIE model principles can only benefit the presentation of a suitable environment, efficiently facilitating students' experience and learning.

8.3   Kodeløypa Making-Based Coding Workshops

"Kodeløypa" is a making-based coding program that consists of workshops that are designed and implemented at the Norwegian University of Science and Technology (NTNU) in Trondheim, Norway. The activities of the workshop are based on the constructionist approach, following the main principles of *making*. The duration of the workshop is approximately four hours and it is conducted in a largely informal setting, as an out-of-school activity. Students from 8 to 17 years old are invited to attend the workshop, which takes place in specially designed rooms, where students work in groups and are introduced to coding and tinkering. Students engage in numerous activities, such as coding digital robots and interacting with them and creating games using Scratch and the Arduino hardware platform. Digital robots are made from recycled materials and an Arduino is attached to each one. Scratch for Arduino (S4A) is an extension of Scratch that provides the extra blocks needed to control the robots. The Scratch programming language uses colorful blocks grouped into categories (motion, looks, sound, pen, control, sensing, operators, and variables), with which children can develop stories, games, and any type of animation (see Fig. 8.1). During the workshop, students work collaboratively in triads or dyads. The design of the workshop also allows students without (or with minimal) previous

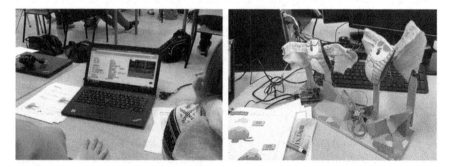

Fig. 8.1 Children creating games using Scratch (left); interactive robots made from recycling materials (right)

experience to attend. Instructors of the activities of the workshop are student assistants, who are responsible for supporting each one of the students' teams as needed. The workshop has two main sessions.

Interacting with the robots: During the first session, the students interact with digital robots. First, one instructor welcomes the students and guides them to be seated, giving a brief overview of the workshop. Each team of students uses one robot. Then with the help of the instructors, students work with a worksheet that is placed on the desks. First, each of the students answers the questions on the worksheet about the exact place and number of sensors and lights on the robots. In addition, students take a tutorial that includes instructions with examples and pictures, similar to the robots they are using. Via the examples shown in the tutorial, students understand exactly how they can interact with the robots. The tasks include the accomplishment of a series of simple loops; those loops will make the robots interact with the environment and perform actions such as turning on a light when sensors detect that the light is below a certain threshold. The students cannot change the different parts of the robots but they can touch and play with them as they want. This section lasts between 45 and 90 min, depending on the team; when everyone finishes the tasks there is a break before the next session.

Creating games using Scratch: This is the main session of the workshop and focuses on the creative implementation of simple game development concepts using Scratch. Students get another paper-based tutorial with examples and visualizations to help them ideate their own game. The tutorial has examples of possible loops that students could use to create their games, including simple text explanations of basic computational thinking concepts. First, the instructors encourage the students to concentrate on discussing ideas for their games and to come up with a draft paper storyboard in collaboration with their team members. Then, again working in teams, students develop their own game by designing and coding using Scratch. To accelerate the children's progress, they are given already existing game characters and easy loops. The instructors support the students while working on their projects, providing help whenever they ask for it. Sometimes, instructors introduce more complex programming concepts on an individual level depending on the needs of their project. Students create their games step by step, by iteratively coding and testing them. In the end, after completing the games, all teams play each other's games. The duration of this session is approximately 3 h.

8.4 Methodology

8.4.1 Focus Group

The study involved five participants: four instructors of Kodeløypa making-based coding workshops and one researcher who participated in focus group sessions discussing how to map those workshops to the instructional model (ADDIE). The researcher's role was to stimulate the brainstorming process and facilitate the sessions with her knowledge, assisting and guiding the instructors' discussions and thinking process. The instructors had a minimum of two years' experience in those workshops and were actively involved in both the instruction and the design decisions. In total, two focus groups were conducted in order to finalize the description of the workshop based on the ADDIE model. During the first focus group session, the researcher presented general information about the existing instructional models, their benefits, and how they are applied and then demonstrated a detailed description of the application of the ADDIE model in different settings. Then, everyone had a clear view of the ADDIE model, its phases, and an overview of how Kodeløypa making-based coding workshops should be investigated and approached in order to be mapped in the model. The aim was to brainstorm ideas and actions in the design of the workshops before and during their execution. Five posters, one representing each of the phases, were hanging on the wall. The task was to use Post-It notes and write down ideas, on an individual level at the beginning, reflecting on each of the five phases of the model.

At each phase of the ADDIE model, instructors spent 15–45 min brainstorming and writing down their ideas; then, they pinned the Post-Its on the respective poster and proceed to the next phase, repeating the same process. At the end of the session and when all the Post-Its were pinned on the posters, the researcher went through all of them and removed the non-relevant ones (if there were any), or asked for more explanations and wrote additional notes if needed. The next session was dedicated to discussing in detail all the ideas that were collected through the Post-It notes, ending up with the most important ones that would describe every aspect of the ADDIE model. Therefore, for each poster (representing the five phases of the ADDIE model) constructive discussions lasted for 30–45 min until there was a general consensus among the participants on the ideas and decisions. The second focus group session lasted for approximately four hours. In the end, the posters with all the ideas were collected by the researcher, who was responsible for organizing the results according to the categories corresponding to the five phases of the ADDIE model: Analysis, Design, Development, Implementation, and Evaluation.

8.4.2 ADDIE Instructional Model Applied to the Kodeløypa Making-Based Coding Workshops—Results

8.4.2.1 Analysis Phase

During the first phase of the ADDIE model, the focus is on analysis and identification of learners in order to determine the instructional goals and learning contexts. More precisely, it identifies the characteristics of the children (i.e., the learners in our case), their existing knowledge, their background, and previous experience, as well as interests and attitudes. Having a clear view of the target audience is important, as it will guide the decisions in the next phases and also provide a realistic approach for the design. Thus, all the available information on the changes in Norwegian reality, including the plans of the Norwegian Government for the schools and educational system, were taken into consideration in how throughout the years Kodeløypa workshops have been evolving and have managed to adjust to circumstances. Three main categories of ideas emerged:

(1) **The learners' background** is the main characteristic that all agreed was the most challenging, as it was very difficult to define in our workshops. In the Kodeløypa workshops, participant children have various backgrounds, as there is an open call to local schools, and no specific prerequisite knowledge from the children is targeted. The main goal is that all children get a general understanding of what coding is and participate in an enjoyable activity outside of the school context, by creating their own projects and collaborating with others. Therefore, the workshop has to be designed in a way that can be adapted to the needs of the children who are participating each time. The background of the children may vary from having zero experience with coding to having a lot and being familiar with more advanced concepts for different reasons; for example, it depends on the school class (if it has coding as an elective subject or a technology class) and each child's individual interest in coding, for instance, trying to code at home or participating in local coding clubs. Consequently, the coding activity has to be adaptable and flexible. The workshop is thus designed for children without (or with minimal) previous experience in coding.

(2) The primary target **age of the participants** is 10th grade; younger or older children can also participate, but each of the workshops should have a specific age group of children, carefully selected regarding age to have the same cognitive capacities. Concerning children's age, the design of the activity (interacting with robots and creating games) and the use of the Scratch programming language (suitable for all ages) provide flexibility and allow for the successful implementation of the workshop with participants from 8 to 17 years old.

As a conclusion to the previous two categories, children who are more knowledgeable in coding can create more advanced games, as the Scratch tool supports it. This is a very challenging process, which all the instructors admitted because they have to have the experience and knowledge to support children in creating their games, from providing very basic to very advanced feedback. Therefore, they all concluded that they should be able to adapt each time depending on the group of children.

(3) A third aspect that emerged was the **gender, motivations, and attitudes** of the children. Most of the time, girls are less exposed to coding than boys and have the impression that coding activities are not interesting for them; this is something that should be taken into consideration, and focus should be given to engaging them in such a way that they think it is not only for boys. Regarding children's motivations and attitudes, attention should be paid to the need to provide a very nice atmosphere during the activity, enhance children's interest in coding and keep even the less-motivated children active in participating in the process.

8.4.2.2 Design Phase

In this phase, the most important aspect for the instructors was to define the learning objectives, but other aspects were also clarified by discussion. Kodeløypa workshops are designed to familiarize children with what coding is and to offer an easy way for them to be introduced to coding by creating their own projects through a pleasant, collaborative activity that lasts for approximately four hours. There are no lectures, but project-based learning methods are applied for high cognitive-level objectives. Instructors have the role of supporting the teams of children depending on their needs and on how they decide to approach the creation of their game, based on their decisions, efforts, and capabilities. Thus, each instructor tries to be in charge of observing two teams. Children working in teams are quite free to act on their own with the instructors as supporters.

The learning objectives of the workshop are implicit, and it turned out that they were never well defined. After the focus group, the following learning goals emerged as expected outcomes from the workshop. The first two categories are connected to coding and problem-solving, the third is related to collaboration, and the fourth to more general benefits and goals:

(1) Learn basic coding skills:

- Learn basic computer science concepts (like loops and variables) and practices (like testing and debugging)
- Be able to create functional code by having an interacting "game".

Using the Scratch programming environment is a good choice, as the basic concepts and practices are well defined, but at the same time "hidden" behind colorful LEGO-like blocks used as commands for children to create their scripts.

(2) Problem-solving skills in game creation and related actions to develop a solution that is new to them by designing and coding a program that meets a set of requirements:

- Investigate the parameters of the problem to guide their approach
- Split the problem into small components
- Generate ideas and alternatives (create their own approach, or explore several possible procedures that might be appropriate to the situation)
- Design a coherent solution
- Test the solution and iterate improvements to satisfy the requirements of the problem.

(3) Collaboration among the children during the process of creating something socially and personally meaningful:

- Decide on the topic that they will start to create
- Share their ideas freely and in a constructive way
- Plan what they have to do when they will do it, and distribute roles and responsibilities if needed

- Discuss issues that occur and give feedback, with the goal to solve problems and be creative
- Make decisions in common for the design of the character and the story.

(4) General:

- Understand the functionality, possibility, and utility of coding environments
- Experience learning but also enjoyment
- Foster a sense of confidence
- Make coding more attractive to girls
- At the least get an understanding of how the creative process in technology happens in order for innovation to take place.

8.4.2.3 Development Phase

In this phase, the outcome represents how the design will be put into action. Below are the two subjects that were discussed and appeared to be important to the instructors:

(1) Together with the **project-based learning method**, the influential aspect of the pedagogical approach is the **"kindergarten approach to learning,"** with the spiral cycle of "imagine, create, play, share and reflect" which is a repeated process (Resnick 2007). One element in this approach that was integrated is "inspiration," which is achieved through a warm-up activity of interacting with the robots and also showing participants similar examples of games. The children think and imagine what they want to create and then they try to make it real. When their games are at an appropriate level to be tested, they share them with the others, reflecting on their experience so far and getting new ideas to continue with their projects. The purpose is for the children to engage in the coding process through exploration, iterations, using different concepts, and trying new elements, with the higher goal of creating the games they want.

(2) The workshop, as described previously, is split into two sessions; it is a largely self-exploratory experience for the children, so the **learning materials** are worksheets and tutorials supporting this process. First, the worksheet is helpful for the children to interact with the robots. It includes questions regarding the position of the sensors, the light-emitting diode (LED) lights, the Arduino board. When it comes to the tutorials, two are needed, one for each session. For the first session, the tutorial helps with the control of the robots using S4A; and the other one, for the second session, aims to support children in the creation of the games and the use of Scratch. The robots tutorial has pictures that are similar to the robots children interact with and gives them simple examples of how to control them with the use of S4A. The second tutorial supports game creation and gives instructions for using Scratch. It starts with an introduction to the Scratch interface and the use of Scratch commands, beginning from the basics, for example, explaining how to set the position of the characters, how to rotate elements, and also providing simple snippets of code for children to try

out. Then, it gives examples of more and more complicated actions, like how to make the characters move, jump, and use collision detection or variables.

8.4.2.4 Implementation Phase

In this phase, the actual delivery of the instruction and the execution of the workshops were discussed. The ideas that emerged relate to what works well, the challenges the instructors are facing, and what they need to focus on in order to effectively and efficiently support the children's learning experience. Therefore, the following aspects appeared to be important:

(1) Usually, the children think they know more than they actually do, so **give them challenges** and **motivate them to use the tutorials**.
(2) **Let the children decide their teams**. Friends collaborate better, as it is not easy to share ideas with someone you do not know.
(3) **The robot part is a good starting point**; everyone participates without problems and uses the tutorial.
(4) **Girls need more support and explanations** because they do not start the activities if they do not have a sufficient understanding of what to do. Also, they read the tutorial more than the boys do.
(5) **Starting to code is the most difficult part** and this is when the instructors should provide the most help and support to the teams. Also, force them to use the tutorial more.
(6) In the case that someone in the teams knows more than others, an option is to motivate him/her to "teach" the other members, rather than having the attitude of creating everything on his/her own to show off; instead, let them all try to be **active participants**.
(7) **Collaboration and discussion** are equally important to other skills and should be enhanced.
(8) **Playing each other's games** is a good motivation for all the children.

8.4.2.5 Evaluation Phase

In this phase, the discussions during the focus group concluded with two main categories. The first refers to ways of assessing the children's learning experience in terms of instruction, how the workshop is designed, and how it is conducted, aiming to get feedback in an ongoing evaluation to improve the activities. The second refers to how to assess the children's learning experience in terms of the learning objectives, their engagement, attitudes, and behavior connected to the research objectives.

For the first category, after the end of the workshop, the children are asked to fill in self-reflection cards individually, where they can anonymously and freely express their thinking about the process and the experience they had. A few questions help the children to elaborate: for example, what they liked most and what they did not like, what they would like to be added to the activity, what they think they have learned, and if they had fun. Also, in the end, a question asks them to write whatever they want and feel it will be useful to share.

For the second category, researchers are responsible for collecting qualitative and quantitative data using various data instruments, including:

- The code the children create in Scratch at different stages, approximately every hour, including the final version.
- Assistants take field notes, conducting structured observations to monitor actions like children's moments of frustration and examples of fun, as well as what kind of help they were getting from the instructors and when.
- Semi-structured interviews with the children at the end of the workshop. The interviews have the purpose of getting as much information as possible from the children on how they experience the making-based coding activity. The questions are related to what difficulties they face during the game creation experience and what is the easiest part, how collaboration is among the members of the team, what frustrates them, and what impresses them.
- Pre–post Scratch evaluation questions. In order to measure the learning gain from their participation in the workshop, the children have to fill in a pre-knowledge acquisition test consisting of coding questions with snippets of code in Scratch, increasing in difficulty, following instructors' suggestions on what the children can acquire from the workshop.
- Using eye-tracking glasses during all parts of the activity, the children's gaze is captured to give insights into their various cognitive mechanisms, predict their progress, and get deeper into their behavior.

All the above-mentioned data have as a higher goal to get a comprehensive view of children's learning experience, extract principles for the design of the workshop, and make further decisions.

8.5   Discussion

This chapter considers how a making-based coding activity, conducted in an informal setting, can be described, mapped, and benefit from an instructional model. In this case, the ADDIE ID model was used in order to discuss the design and development of a learning experience. During two focus group sessions with the instructors of the making-based coding workshop, we discussed the development of the workshop based on the model's five phases; after the sessions, the most important aspects were revealed, concerning what to think about in the design of coding workshops when applying the ADDIE model. In addition, we supported the fact that it is possible to fit an activity outside of formal settings into an instructional model that has not applied it before and benefit from its systematic approach.

During the focus group, from the researcher's point of view, who was also the facilitator, it was difficult sometimes to guide the discussions. The instructors' final decisions, as Post-It notes and ideas, turned out to be more intuitive and not expressed properly. This is due to the fact that they were not familiarized with the model's five phases and what exactly is needed to be addressed in each of them. In particular, the discussions in the analysis and design phase were the most challenging; on the other hand, from all the five phases, the one with the most effective discussions was the implementation phase. When the workshop was initially designed, the focus was to familiarize the students with coding, show them the possibilities of a programming environment, help them become aware that they can be creators rather than simply consumers of technology, and overall give them an idea about computer science. This makes it difficult to determine exactly the identity of the learners, because the characteristics of the possible participants of the workshop are very broad, regarding both age and background. One solution is to design flexible and adaptable activities (Papavlasopoulou et al. 2019). However, based on their experience, in the analysis phase of the ADDIE model, the instructors managed to focus on the most prominent characteristics of the learners, like age, background, gender, motivation, attitudes, and so on. In their study, Ozdilek and Robeck (2009), analyzing the responses of instructional designers in various areas of education, found that the analysis step of the ADDIE model was the most challenging and that most of the attention is given to learner characteristics compared to other elements. One of the important aspects that was shown is the importance of designing an enjoyable activity, which is in line with similar workshops (Norouzi et al. 2019).

Regarding the learning objectives that were specified in the design phase, they were apparently in the instructors' minds, but it was difficult for them to express and explain what they were thinking about exactly before taking part in the focus group. After the discussions, the learning goals were clear and well defined by everyone, placing them on common ground; it is apparent that the focus is not only on learning coding but also on the overall experience. Furthermore, the learning materials have to be in line with the learning goals and support the smooth execution of the workshop (Liu et al. 2008); their development has to be carefully and strongly connected with the design of the workshop. Novák et al. (2018) used the five phases of the ADDIE model as a strategy to develop educational materials for the use of the Arduino platform. During the discussions for the implementation phase, it was obvious that instructors were more active and efficient, without getting much help from the researcher to explain and guide them; discussing the execution of the workshops, needs, and challenges was something more natural to them. However, the implementation phase needs constant revision based also on the results of the evaluation, which requires an appropriate approach. For example, researchers and instructors should agree on the evaluation strategy, and then the researchers should communicate the results to the instructors. In this way, they will introduce a teaching approach and design decisions, with a higher goal of creating a beneficial learning experience for the children.

In general, despite some challenges, the instructors found the ADDIE model really interesting and very helpful for understanding, framing, and advancing the design of the Kodeløypa making-based workshops. In the focus groups, important aspects were revealed of what to consider as the main characteristics in order to develop a similar workshop, indicating how it has to be approached: for example, have clear learning objectives; consider the most important aspects of the learners' identity; during the activities act accordingly, like motivating them to use the tutorials more, to get help, and to support boys' and girls' teams depending on their needs and capabilities. In addition, some of the discussions and ideas described have not been explicitly implemented in the Kodeløypa workshops in their current state but examining the use of the instructional model and how it is implemented gave instructors the opportunity to think about future improvements and plans, like deciding on the use of appropriate evaluation techniques and how to implement the workshops more effectively, with the correct choices based on the circumstances and the characteristics of the learners. Reflections on the basic structure of the workshop also helped the instructors to see things more clearly and develop more ideas, despite the fact that these were not expressed or were explicit from the beginning of the focus group. This indicates that the ID process gives the instructors the opportunity to think about and understand the steps and the process of how to follow a specific model and theory, consequently leading to a better design of the learning activity and helping them to become better (Khalil and Elkhider 2016).

Future work should focus on adjusting the model, such as adding and reassuming the phases of the model, depending on the needs of the learning activity that has to be developed. This will allow a better-designed experience for the learners and customize the needs of the instructors or other stakeholders who are interested each time. This study is limited in that the ADDIE model is used in one case; we suggest the need for confirmation in other similar cases, which will show evidence and contribute to the use of the ADDIE design model to inform, guide, and lead to successful educational experiences (Smith and Boling 2009). However, we should maintain a critical point of view, and not forget the limitations of the model and the fact that it has been criticized as not always being very effective (Bichelmeyer 2004).

Acknowledgements This work is supported by the "Learning science the fun and creative way: coding, making, and play as vehicles for informal science learning in the 21st century" Project, under the European Commission's Horizon 2020 SwafS-11-2017 Program (Project Number: 787476) and the "Learn to Machine Learn" (LearnML) project, under the Erasmus+Strategic Partnership program (Project Number: 2019-1-MT01-KA201-051220).

References

Alley, L. R., & Jansak, K. E. (2001). The ten keys to quality assurance and assessment in online learning. *Journal of Interactive Instruction Development, 13*(3), 3–18.

Bichelmeyer, B. (2004). The ADDIE model—a metaphor for the lack of clarity in the field of IDT. AECT 2004 IDT Futures group presentations IDT Record.

Hall, B. (1997). *The web-based training cookbook with CDRom.* Hoboken, NJ: Wiley.

Hidayanto, D. R., Rahman, E. F., & Kusnendar, J. (2017). The application of ADDIE model in developing adventure game-based multimedia learning to improve students' understanding of basic programming. In *2017 3rd International Conference on Science in Information Technology (ICSITec)*, pp. 307–312. IEEE.

Khalil, M. K., & Elkhider, I. A. (2016). Applying learning theories and instructional design models for effective instruction. *Advances in Physiology Education, 40*(2), 147–156.

Liu, E. Z. F., Kou, C. H., Lin, C. H., Cheng, S. S., & Chen, W. T. (2008). Developing multimedia instructional material for robotics education. *WSEAS Transactions on Communications, 7*(11), 1102–1111.

Morrison, G. R., Ross, S. J., Morrison, J. R., & Kalman, H. K. (2019). *Designing effective instruction* (p. 480). Chichester: Wiley.

Norouzi, B., Kinnula, M., & Iivari, N. (2019). Interaction order and historical body shaping children's making projects—a literature review. *Multimodal Technologies and Interaction, 3*(4), 71.

Novák, M., Kalová, J., & Pech, J. (2018). Use of the Arduino platform in teaching programming. In *2018 IV International Conference on Information Technologies in Engineering Education (Inforino)*, pp. 1–4. IEEE.

Oliveira, G. A. S., & Bonacin, R. (2018). A method for teaching object-oriented programming with digital modeling. In *2018 IEEE 18th International Conference on Advanced Learning Technologies (ICALT)*, pp. 233–237. IEEE.

Ozdilek, Z., & Robeck, E. (2009). Operational priorities of instructional designers analyzed within the steps of the Addie instructional design model. *Procedia-Social and Behavioral Sciences, 1*(1), 2046–2050.

Papavlasopoulou, S., Giannakos, M. N., & Jaccheri, L. (2019). Exploring children's learning experience in constructionism-based coding activities through design-based research. *Computers in Human Behavior, 99*, 415–427.

Pribadi, B. A. (2009). *Model desain sistem pembelajaran* (p. 35). Jakarta: Dian Rakyat.

Resnick, M. (2007). All I really need to know (about creative thinking) I learned (by studying how children learn) in kindergarten. In *Proceedings of the 6th ACM SIGCHI conference on Creativity & cognition*, pp. 1–6.

Salahli, M. A., Yildirim, E., Gasimzadeh, T., Alasgarova, F., & Guliyev, A. (2017). One mobile application for the development of programming skills of secondary school students. *Procedia Computer Science, 120*, 502–508.

Smith, K. M., & Boling, E. (2009). What do we make of design? Design as a concept in educational technology. *Educational Technology, 49*(4), 3–17.

Wu, P. (2014). A game programming instructional design model. *Journal of Computing Sciences in Colleges, 29*(6), 57–67.

Sofia Papavlasopoulou is a post-doc researcher at the Department of Computer Science, Norwegian University of Science and Technology (NTNU). She holds a Ph.D. from NTNU in the area of Learning Technologies. Her interests center on the use of technological tools to support students' learning while enhancing their interest in Computing Education and coding, providing creative engagement. Her goal is to investigate the best ways to support an interactive, engaging approach to informal learning coding activities for young students and design meaningful learning experiences for them. Since 2015, she has co-authored more than twenty-five research articles in peer-reviewed journals and conferences. The results of her studies have been published in leading Journals like Computers in Human Behavior (Elsevier) and Entertainment Computing (Elsevier), International Journal of Child-Computer Interaction (Elsevier), and conferences like the ACM Conference on Interaction Design and Children (IDC).

Michail Giannakos is a Professor of interaction design and learning technologies at the Department of Computer Science of NTNU, and Head of the Learner-Computer Interaction lab (https://lci.idi.ntnu.no/). His research focuses on the design and study of emerging technologies in online and hybrid education settings, and their connections to student and instructor experiences and practices. Giannakos has co-authored more than 150 manuscripts published in peer-reviewed journals and conferences (including Computers & Education, Computers in Human Behavior, IEEE

TLT, Behaviour & Information Technology, BJET, ACM TOCE, CSCL, Interact, C&C, IDC to mention few) and has served as an evaluator for the EC and the US-NSF. He has served/serves in various organization committees (e.g., general chair, associate chair), program committees as well as editor and guest editor on highly recognized journals (e.g., BJET, Computers in Human Behavior, IEEE TOE, IEEE TLT, ACM TOCE). He has worked at several research projects funded by diverse sources like the EC, Microsoft Research, The Research Council of Norway (RCN), US-NSF, German agency for international academic cooperation (DAAD) and Cheng Endowment; Giannakos is also a recipient of a Marie Curie/ERCIM fellowship, the Norwegian Young Research Talent award and he is one of the outstanding academic fellows of NTNU (2017–2021).

Chapter 9
Guidelines for Empowering Children to Make and Shape Digital Technology—Case Fab Lab Oulu

Marianne Kinnula, Netta Iivari, Iván Sánchez Milara, and Jani Ylioja

Abstract Digital technology design and making skills are seen as important 'twenty-first century skills' that children need to learn to become future changemakers, i.e., to manage and master in the current and future technology-rich everyday life. Fab labs (digital fabrication laboratories) are one example of non-formal learning environments where schoolteachers bring children to work with projects on digital technology design and making. Even though the value of fab labs in such endeavors has been acknowledged, the potential of fab labs in empowering children to make and shape digital technology remains poorly explored. This study scrutinizes the current theoretical understanding of empowerment related to design and making and relates that on empirical data of practical work done with children in the fab lab of the University of Oulu. Based on that, we offer theory- and practice-based guidelines for practitioners who wish to empower children to make and shape digital technology in the context of non-formal learning and fab labs. These guidelines should be useful for teachers when planning and implementing children's work in fab labs as well as for fab lab personnel who help children to conduct their projects, with special emphasis on school visits to fab lab premises.

Keywords Fab lab · Digital fabrication · Making · Non-formal learning · Empowerment · School · Children · Pupils · Teachers · Facilitators · Instructors

M. Kinnula (✉) · N. Iivari
INTERACT Research Unit, University of Oulu, Oulu, Finland
e-mail: marianne.kinnula@oulu.fi

N. Iivari
e-mail: netta.iivari@oulu.fi

I. Sánchez Milara
UBICOMP Research Unit and Fab Lab Oulu, University of Oulu, Oulu, Finland
e-mail: ivan.sanchez@oulu.fi

J. Ylioja
Fab Lab Oulu, University of Oulu, Oulu, Finland
e-mail: jani.ylioja@oulu.fi

© Springer Nature Singapore Pte Ltd. 2020
M. Giannakos (ed.), *Non-Formal and Informal Science Learning in the ICT Era*, Lecture Notes in Educational Technology,
https://doi.org/10.1007/978-981-15-6747-6_9

9.1 Introduction

Our society and everyday life are becoming extensively permeated by digital technologies, and it is hard to see many future occupations that have not been affected by digitalization. Access to technology and ability to benefit from its use (Iivari et al. 2018b; OECD 2012; Warschauer 2002), as well as skills and capabilities to innovate, design, program, make, and build digital technology (Blikstein 2013; Iivari and Kinnula 2018; Iivari et al. 2016; Iversen et al. 2017), are all seen as pivotal for children to manage and master in the current and future technology-rich everyday life. Children's education needs to respond to this—children's education should empower them to make and shape digital technology in addition to using it in meaningful ways (see e.g., Iivari and Kinnula 2018; Iivari et al. 2016; Iversen et al. 2017).

Various kinds of actions and developments have already emerged around the topic. Schools and teachers around the globe are facing the challenge of educating children to meet these needs of the future digitalized society and workforce. Even if impressive developments have taken place, such as the FabLab@school.dk project (Iversen et al. 2018), there still is an acute need to develop children's education in this respect (e.g., Kinnula et al. 2015; Smith et al. 2015). In addition to schools and teachers, a significant work has been undertaken in the informal and non-formal learning (see e.g. Eshach 2007) for the differentiation) contexts: Makerspaces, fab labs, and different kinds of computer or coding clubs have started to offer children digital technology skills and competencies (e.g., Bar-El and Zuckerman 2016; Chu et al. 2015; Katterfeldt et al. 2015; Litts 2015; Posch et al. 2010; Weibert and Schubert 2010). However, their potential in empowering children to make and shape digital technology remains poorly explored so far.

This study addresses particularly fab labs as a promising site to offer children digital technology skills and competencies. Fab labs ((Digital) Fabrication Laboratories) are communal, small-scale digital fabrication and innovation platforms with a mission to popularize processes of turning something digital to a physical object or a functional device. They originate from MIT's outreach program (Gershenfeld 2012). There are certain criteria for fab labs, set by the Fab Foundation[1]: (1) a public access to the fab lab; (2) the fab lab subscribes to the fab lab charter, a basic rule set for all fab labs; (3) it shares a common set of tools and processes with other fab labs; and, (4) it participates in the larger, global fab lab network. While fab labs are paving the way for the third digital revolution of digital to physical and ubiquitous fabrication of programmable materials, the value of fab labs comes more of learning the processes than of the actual outcome of the processes (Gershenfeld et al. 2017). Fab labs are sometimes described as places to learn, mentor, play, create, and innovate, not to forget communal co-working.

In this paper, we ask as our research questions: What is the potential of fab labs in empowering children to make and shape digital technology? What kind of best practices, limitations, or challenges can be identified? Inspired by a study by Kinnula and Iivari (2019), we address the empowerment of children to make and shape digital

---

[1] https://www.fabfoundation.org.

technology by relying on a framework proposed by Chawla and Heft (2002) that considers how to enable children's effective, genuine participation in projects of various kinds. We apply the framework in the context of fab labs. To answer our research question, we examine data from a collaborative workshop conducted with the University of Oulu Fab Lab personnel who regularly work with children. In the workshop, we collaboratively discussed the framework on the empowerment of children and reflected on how it has been realized in the fab lab when working with children. Based on the insights generated, we develop guidelines for practitioners working with children and their digital technology education in the context of non-formal learning and fab labs. These guidelines should be useful both for teachers and facilitators when planning and implementing children's projects in fab labs, with special emphasis on school visits to fab lab premises.

The paper is structured as follows: The next section introduces the theoretical background and related research. Then we describe our research methods and present our findings from our data analysis. After that, we propose practical guidelines on what kind of aspects different actors working with children in the fab lab environment should consider when they aim at empowering children with digital technology. In the last section, we conclude the paper and discuss limitations and future research possibilities.

9.2   Empowering Children with Digital Technology

Empowerment is a complex concept discussed within numerous disciplines, and, as a result, there is a variety of meanings associated with it (Kinnula et al. 2017). In this paper, by the empowerment of children to make and shape digital technology, we mean children's perceived competence, impact, meaningfulness, and choice around making and shaping digital technology (Spreitzer 1995; Thomas and Velthouse 1990); we acknowledge, however, that there are other views of empowerment of children in this context as well; see e.g., Kinnula and Iivari (2019) and Kinnula et al. (2017).

Children's empowerment has been considered in different fields and for different purposes. In this paper, we rely on the framework by Chawla and Heft (2002), following the study by Kinnula and Iivari (2019). Chawla and Heft (2002) outline a number of criteria for genuine, effective participation of children (see Table 9.1). In their framework, they highlight the following aspects: (1) children's participation should be meaningful to participating children; (2) their participation should have an actual impact on the results; (3) their participation should lead to their competence building; and (4) children should have a choice to decide whether they participate or not. The criteria are grouped under five conditions that need to be respected when working with children. These criteria align very well with literature on empowerment ((Thomas and Velthouse, 1990), see also Spreitzer 1995) and hence we interpret them as criteria for the empowerment of children.

Table 9.1 Criteria for the empowerment of children (Chawla and Heft 2002, p. 204)

Conditions of convergence	Conditions for competence
Whenever possible, the project builds on existing community organizations and structures that support children's participation As much as possible, project activities make children's participation appear to be a natural part of the setting The project is based on children's own issues and interests	Children have real responsibility and influence Children understand and have a part in defining the goals of the activity Children play a role in decision-making and accomplishing goals, with access to the information they need to make informed decisions Children are helped to construct and express their views
Conditions of entry	There is a fair sharing of opportunities to contribute and be heard
Participants are fairly selected Children and their families give informed consent Children can freely choose to participate or decline The project is accessible in scheduling and location	The project creates occasions for the gradual development of competence The project sets up processes to support children's engagement in issues they initiate themselves The project results in tangible outcomes
Conditions of social support	*Conditions for reflection*
Children are respected as human beings with essential worth and dignity There is mutual respect among participants Children support and encourage each other	There is transparency at all stages of decision-making Children understand the reasons for outcomes There are opportunities for critical reflection There are opportunities for evaluation at both group and individual levels Participants deliberately negotiate differences in power

In their paper, Kinnula and Iivari (2019) critically consider these criteria and propose a set of questions to ask when aiming at empowering children to make and shape digital technology. However, the questions have not been empirically evaluated with practitioners. That is where this study contributes.

The topic of empowerment of children to make and shape digital technology has been acknowledged as significant in Child–Computer Interaction (CCI) research long ago (see Read and Markopoulos 2013). Druin and her colleagues have already for decades argued for including children into the technology design process as design partners, i.e., as equal participants to adults, having valuable expertise on what being a kid entails that should be utilizable in the design process (e.g., Druin 2002; Druin et al. 1999). Druin's work has been inspired by the Scandinavian tradition of Participatory Design and, along these lines, many researchers have developed the ideas further. For example, Iversen and colleagues (2017) have argued for inviting children as protagonists in relation to technology: children are to critically reflect on technology and its role in our everyday life and practices as well as to take an active role in shaping technology development. Iivari and Kinnula (2018), along these lines, have considered what such protagonist role adoption entails. They, additionally, have considered how well the context of school enables genuine participation of children (Iivari and Kinnula 2016) as well as the ways by which they have managed to empower children in technology design (Kinnula et al. 2017). Discussion on the empowerment of children strongly relates to ethical issues and values driving or underlying our work with children. Indeed, CCI research has already brought up ethical issues and the importance of values shaping design work with children (e.g., Iversen et al. 2010; Iversen and Smith 2012; Kinnula and Iivari 2019; Kinnula et al. 2018; Read et al. 2014; Read et al. 2017; Read et al. 2013; Van Mechelen et al. 2014).

This study will particularly scrutinize fab labs as a site for design and making projects. Studies have already examined children's design and making activities in fab labs (Blikstein and Krannich 2013; Iivari and Kinnula 2018; Iivari et al. 2018a; Iversen et al. 2016; Katterfeldt et al. 2015; Posch and Fitzpatrick 2012; Posch et al. 2010; Pucci and Mulder 2015). However, the specifics of fab labs in helping or hindering the empowerment of children as regards digital technology have not been explored so far. This study provides valuable insights into the particularities of fab labs as such a site.

9.3 Methods

This exploratory study has been conducted in Finland, in the context of the University of Oulu Fab Lab ("Fab Lab Oulu"), which is equipped with a large toolset (e.g., a laser cutter, a sign cutter, a precision Computerized Numerical Control (CNC) milling machine, a large-scale router-type milling machine, 3D printers, a computerized embroidery machine, computers, electronic workstations, and communication devices for videoconference). Working in a fab lab supports distributed education and knowledge sharing (Ylioja et al. 2019). Most of the processes are easy enough for almost anyone to learn, and the majority of fab labs in Nordic countries, and also in Oulu area, are in schools and educational institutes. Fab Lab Oulu is open to everybody but one of its core goals is to get the local school community familiar with it to create a community of school teachers, pupils, and university researchers around the fab lab, and attract new students to the technical faculties of the university.

Regarding community building, Fab Lab Oulu started collaboration with the local schools right after it opened with the model of acting as a 'super node' for the schools of the area. The aim is to familiarize teachers and pupils with basic processes and support their projects with the equipment set of Fab Lab Oulu. This model permitted the local schools to start their own makerspaces, where they can do a large part of their making projects. When the resources of the schools are not adequate to finalize the project, they can visit the super node. Regarding attracting new students, we believe that seeing a tangible outcome of one's work is highly motivating. We strongly believe that becoming familiar with simple manufacturing tools and processes, throughout the realization of different design and making activities, makes it more likely that pupils consider related studies as one possible option for their career choice.

For the purposes of this study, two of the authors conducted two workshops involving five researchers somehow working with Fab Lab Oulu, with varying overlapping backgrounds (Cultural Anthropology/Information Systems/Human–Computer Interaction/Computer Science Engineering/Electrical Engineering) and roles (fab lab director/fab lab manager/fab lab instructor/coordinator for education-related activities in the fab lab/part-time researcher/researcher organizing research projects in collaboration with local schools and the fab lab). The researchers altogether have over 30 years of experience working with children. Four of the participating researchers are the authors of this paper.

The first workshop pre-reading material included a framework on the empowerment of children (see Kinnula and Iivari 2019) as well as definitions of formal/non-formal/informal learning. In the 3-hour workshop, we followed an agenda based on different parts of the framework. The focus was on the framework's conditions for meaningful and impactful participation. We collaboratively reflected on the current practices, goals, and values of Fab Lab Oulu and its personnel in relation to the framework on the empowerment of children to understand how it has been realized in Fab Lab Oulu when working with children, asking ourselves the reflective questions provided in the framework. All five researchers participated in this workshop. The workshop was audiotaped and one of the participants wrote down extensive, reflective notes of the discussions. The notes were shared between the participants.

Between the two workshops, data analysis continued. First, one of the researchers sorted out the discussion content to extract main insights from the framework on the empowerment of children and shared this with the rest of the research team. Afterward, the research team members individually concretized, clarified, and extended the insights. This collaborative process resulted in a summary, addressing all conditions in the empowerment framework, proposing aspects for all fab labs to follow when preparing activities with schoolchildren. Those insights were then used to define a set of guidelines, divided into different phases that could be useful for anyone trying to organize any kind of educational activity for schools in the fab lab environment.

In the second workshop (1.5 h), three of the participants collaboratively discussed the results gained and reflected on them critically, and further refined their presentation and relation to each other.

9.4 Findings

In the following sections, we discuss our findings on how the framework on the empowerment of children has been realized in the work of Fab Lab Oulu. Our findings are structured according to the conditions presented by Chawla and Heft (2002) and related reflective questions, presented by Kinnula and Iivari (2019). Insights based on our findings are presented later in the discussion section of the paper as practical guidelines on what kind of aspects different actors working with children in the fab lab environment should consider when they aim at empowering children with digital technology. Before moving on to the findings, it needs to be noted that we recognized that Fab Lab Oulu has organized multiple types of educational activities involving children of different ages, with different goals and formats. We realized that the goal and format of the activity are important when studying children's empowerment using the given framework: different things as regards the empowerment of children become emphasized and realized in different types of educational activities. We divide the activities into three types, each having implications on the empowerment of children, discussed later on.

Short-term activities. These are usually about 2-h-long school visits or even shorter, aiming to give pupils a basic understanding of what happens in the digital fabrication process: first, they design a simple object with a computer (a piece of jewelry, sign for their room, a keychain or such, using a 2D vector graphics software) and then use a machine to fabricate a physical object (in this case, by using the laser cutter). We provide a guided tutorial, explaining step by step and with real-time support how to utilize different software functionalities to build the object. Children must follow the given steps. They do not do significant creative work; all designs look similar. Finally, we cut the design by using the laser cutter. We show the cutting process to children, but they do not intervene in the full process. They perform some simple actions such as setting the origin or starting the machine. Children work individually or in pairs.

One-day activity. During this activity, pupils learn the basics of one concrete digital fabrication process, for instance, laser cutting, vinyl cutting, 3D design, or introduction to electronics. The activity starts with a guided tutorial on how to use the software tool but, afterward, pupils must work on their own project. The instructor usually defines a problem or certain functional/design conditions that the project must implement. Children work on their own with the support of instructors. During the fabrication process, children are taught how to use the machines. If they need to go through a second or third iteration, they can use the machines on their own. Work in small teams is encouraged.

Longer term activity. In this category, children typically work in 5–10-day-long projects where they must use different digital fabrication processes. In case of a school group, the teacher usually sets requirements for the project. In other cases, the fab lab instructors set the requirements for the project. Children work with total freedom and they need to make their own decisions. Instructors, acting mainly as facilitators, provide support to children when needed; they explain processes and point children to online material and tutorials for more information. Outcome of the project is a tangible object that involves the usage of different digital fabrication processes at the same time. Usually, children work in teams of several members.

9.4.1 Conditions of Convergence

The conditions of convergence emphasize that one should utilize aspects from children's existing life world as much as possible when aiming at empowering them. In our workshop, several interesting observations were raised around this.

The first reflective question in the framework on the empowerment of children asks, **is it easy and natural for children to participate?** The local instructors agree that Fab Lab Oulu is not the type of place that the average child is familiar with: It is located at the university campus and for children visiting our fab lab for the first time, it is commonly the first time they visit university premises as well. It is also a space full of people working on their own projects and with lots of unfamiliar noisy machines. The activities carried out at the fab lab might look somewhat chaotic to children.

Differences between the structures of activities can also cause confusion to the pupils: Most activities carried out at fab lab are semi-structured or totally unstructured (with the exception of short-term activities). This can be a radical change for children who are familiar with more structured activities in their classes. Sometimes they might feel a little bit lost, without knowing how to continue. In general, our fab lab instructors observe that children, no matter the age, seem to be used to much guided activities and find it difficult to search for information by themselves. Hence, it is important for instructors to follow the general atmosphere and act when there is a clear decrease in engagement. This can be challenging to the instructors who often lack the pedagogical background and do not know the children, their strengths and weaknesses.

Despite these differences, we have noticed that pupils often adapt quite easily to the new environment. Especially the older pupils (aged 13–18) seem to value the freedom of the work. The new Finnish National Curriculum for education (NCBE 2014) also promotes project-based learning. Hence, pupils are becoming more familiar with the fab lab learning style. Thanks to the dissemination work made at Fab Lab Oulu and the growing interest of the local school community in STEAM pedagogies and digital fabrication (Sánchez Milara et al. 2019), the schools are also gradually integrating digital fabrication into formal education as another educational technology asset. Some of them have their own makerspaces as well. Thus, many of the pupils from those schools are already familiar with the basic concepts, processes, and machines. For those children, the fab lab is not as intimidating environment as might be for those who come to a fab lab for first time.

Even though pupils get familiar with the fab lab in their school projects, our instructors have noticed that children very rarely return to the fab lab to work on their own project. We have not explored yet the reasons, but as pupils visit the fab lab as part of their schoolwork, we assume that they associate the fab lab with school and not with an activity that they can do during their free time. So, organizing voluntary activities outside the school environment might help children see digital fabrication also as a hobby.

The second question regarding conditions of convergence is, **are existing organizations and structures supporting children's participation relied on?** Fab Lab Oulu seldom organizes activities for children if a learning institution has not applied for it beforehand. Schools usually initiate the process by informing the fab lab staff that they would like to visit and giving information about group size and age of participants. During 2018, Fab Lab Oulu hosted around 1000 5–18-year-old pupils.

One aspect that our remark is that teachers stay at the fab lab during the activities but act just as observers and rarely intervene. It seems that teachers feel they do not master the topic, and hence they should step back. The instructors, then again, like to keep their facilitator or instructor role. In line with Laru and colleagues (2019) and Pitkänen and colleagues (2019), we believe that when working in a fab lab is part of school activity teachers should be the ones designing the learning activity, including learning outcomes, didactics, and pedagogies. To that end, teachers should become aware of the fab lab potential and have certain knowledge of the processes. Fab Lab Oulu is currently training teachers both in the technical aspects and in the pedagogical use of processes and machines of the fab lab. Our instructors believe that to promote more educational activities in a fab lab and to integrate them into formal education, the activities should have a clear learning outcome that matches the ones in the National Curriculum.

In addition to school activities, building a community around a fab lab is important and Fab Lab Oulu has been doing this in many ways. A local startup, run by university students, organizes after school clubs for children over 11-year-old in the fab lab premises. In addition to that, they bring a Mobile Fab Lab, i.e., a small set of fab lab machines packed in a van, to schools all around Finland with the goal of presenting the fab lab concept and digital fabrication to pupils and teachers. During 2018, they reached 2000 students. Fab Lab Oulu also attends different events organized by other institutions (especially events concerning education). Furthermore, in events organized by the university, we advertise Fab Lab Oulu to the general public. Fab Lab has participated in the European wide Researchers' Night, e.g., for 3 years in a row. In 2019, more than 1400 visitors were hosted in just 4 h. We believe that this is necessary to increase awareness of the fab lab, to show education stakeholders and parents the potential of fab lab for children as well as to create synergies with other organizations. Fab Lab Oulu also aims to attract current pupils to become future students in our university. To that end, collaboration with local schools is important and teachers should consider themselves as part of the community as well (Sánchez Milara et al. 2019). Creating a community takes a lot of time and effort but a fab lab can be only successful if there is a strong community around it.

Finally, concerning existing structures, we think that social media is an important source to advertise fab lab potential and different events. Fab Lab Oulu uploads the most meaningful project results in different social media (Facebook, Twitter, Instagram, and Vimeo) and encourages pupils to share their work. We assume that this would make their own work more meaningful to themselves.

The third question regarding conditions of convergence is, **are the activities based on children's own issues and interests?** Actually, the entire fab lab concept originated from the interest of people to learn how to build objects that they cannot buy at shops (Gershenfeld 2012). However, before it is possible to realize what can be done at a fab lab and to start working in fabricating something meaningful you need to learn the basic functionalities and processes. When answering this question, time constraint is a central issue: Very short-term activities (~2 h) in Fab Lab Oulu are guided activities, with the goal that children learn basic principles of a process and instructors decide the object children are fabricating and the process they need to use. The activity should still, of course, lead to a purposeful object with a certain level or personalization for the participating children. In that sense, short-term activities are not based on children's own issues or interests. In longer activities, the design process is part of the learning process and pupils have a lot of freedom to decide on what to work with.

9.4.2 Conditions of Entry

The conditions of entry emphasize broad, inclusive, voluntary, and accessible participation of children. In our workshop, interesting observations emerged.

The first questions asked, **Have the participants been fairly selected? Has somebody been excluded? Why?** Short-term and one-day activities at Fab Lab Oulu premises are usually initiated by teachers who would like their pupils to visit our premises. They get to know fab lab educational activities either through word of mouth, through advertising made by city representatives, or thanks to any of our other activities (e.g., Mobile Fab Lab or Researchers' Night). Thus, teachers decide which classes and pupils visit the fab lab. In order to warrantee equality of opportunities among all schools in the area (including schools located tens of kilometers away from the fab lab premises and with no good access through public transport), the city government funds bus tickets for all classes who desire to visit the fab lab. In addition, one-day Mobile Fab Lab visits are arranged for those schools far away from Fab Lab Oulu premises. All schools in the area have been informed of the existence of this service. Thanks to these two measures, all local schools have had the chance of experimenting with digital fabrication.

We advertise the long-term activities among all local schools, by social media, by emailing to school administration, and by local city channels. After that, we open the registration process. If there are more applicants than open positions, we select the participants based on a motivation letter and CV/course grades.

The inclusion of students with special needs was raised during the workshop discussions. Among Fab Lab Oulu staff there are no experts to deal with special needs, but all children are welcome to the fab lab to carry out activities, and we have hosted pupils with different kinds of behavior and social disorders (e.g., autism spectrum disorder, ASD) with support from their teachers. Other than a computer room on the second floor, the premises are accessible for wheelchair users. With pupils with motor disabilities, it is possible to use laptops downstairs and the staff has arranged videoconferencing system so that people downstairs can follow any class upstairs. All fab lab education involves group work. In that way, even when not all fab lab machines can be used by people with disabilities, children can participate in the majority of tasks. Their team members might perform the tasks that the person with disabilities cannot do.

The next questions regarding conditions of entry asked, **is children's participation voluntary? Why? If not, why?** Nature of the activity affects this in Fab Lab Oulu. If the activity is part of school work pupils' participation is not voluntary; it is part of their formal education and thus mandatory for them. As any mandatory task, some children like it and some do not, although mostly they seem to enjoy working in the fab lab. All activities not organized by schools are voluntary for children. Only children interested in the activity register themselves. Sometimes, they are influenced by teachers or parents, but the instructors have never noticed a child that seemed to be forced to attend.

The next question regarding conditions of entry asked, **is the location and schedule for the activities easy to access for children and their families?** Fab Lab Oulu is located next to the university's main entrance. The campus is located 6 km away from the city center. University is well connected with the city either by public transport or by car. In addition, several cycling lanes cross the university campus. Most pupils from local schools come to fab lab premises either by biking or by public transport. Schools further away usually rent a bus. The majority of pupils participating in activities outside school hours come by bike or by public transport. We believe that the location is good. The current location inside university premises feels like a good selling point to attract children to get to know the university and get some exposure to university life. Children participating in the activity usually share space with university students (in either the fab lab or other common places such as canteen). In addition, a fab lab visit is a good opportunity to get to know some of the research activities carried out at the university. For instance, many of the pupils attend a talk on virtual reality given by researchers expert on the topic.

Concerning the opening times, Fab Lab Oulu is always open during school time making school visits possible. In addition, the instructors agree that it is also necessary to have the fab lab open after school. In that way, children can later visit the fab lab on their own to work with their own projects or attend workshops organized at the fab lab.

9.4.3   Conditions of Social Support

The conditions of social support aim for a supportive and encouraging atmosphere. In order to analyze the findings, we have divided the related questions into two different sets: collaboration in fab lab and behavior of participants.

The questions related to collaboration asked, **Is the environment supportive? Is there a team spirit? How can this be encouraged? Do children support and encourage each other? How can this be supported? Are everybody's opinions and thoughts considered valuable?** Collaboration is one of the horizontal competencies emphasized in the majority of twenty-first-century skills frameworks (e.g., the one proposed by Ananiadou and Claro 2009). Maker culture has collaboration and sharing of knowledge as one of its foundations (see e.g., Dougherty 2016). We believe that working at fab lab can be a natural way to learn collaboration and our instructors usually suggest pupils finishing their task earlier to, e.g., help other pupils. Particularly in longer term activities, collaborative aspects become more visible as the activities are not very guided and usually some distribution of tasks between group members is needed to succeed in the project. We believe that the blurred role of teachers in Fab Lab Oulu (usually acting more like facilitators than traditional teachers), as well as instructors' expert role, facilitate collaboration among children as they face that they should solve the problems by themselves without a teacher constantly guiding the steps.

One possible occasion to show the value of everybody's thoughts and opinions is when reflecting on the work done. In Fab Lab Oulu, instructors usually finalize each session of longer activities (one-day/long-term activities) with a short reflection on the status of the activities and problems faced. During this reflection, all children are encouraged to give their view. The idea is to share possibly useful approaches and show that similar problems occur. It is possible to learn from others how those can be solved. We have noticed, however, that it is very difficult to get all children to talk, and hence the efficacy of the reflection session might be smaller.

The second group of questions related to conditions of support asked: **Are all participants respected? Do all participants act friendly and politely? How can this be encouraged**? This is not an aspect that Fab Lab Oulu staff has worked explicitly, and they don't have training for how to handle, e.g., behavioral problems with children. With school groups, teachers' role is important, as they know the children and the type of relationship they have established with each other.

9.4.4   Conditions for Reflection

The conditions for reflection aim for increased transparency in the process: who makes decisions, why these decisions are made, and how the participants evaluate the results. The first questions asked, **do power differences exist between participants? Have the power differences been deliberately negotiated?** In Fab Lab Oulu, power differences are completely different from a normal school setting as the role of teachers and instructors are blurred. We think that in our short-term activities, the instructors act more like digital fabrication teachers, giving instructions that participants must follow. However, in our long-term activities, instructors act as experts in digital fabrication as well as facilitators, giving only pointers for children when they do not know how to continue. Our instructors argue that teachers' role changes in fab lab environment; they need to act more like counselors than teachers. In that sense, the power differences are clearly reduced. Between children, due to differences in personality and skills, they spontaneously take different roles: some of them become group leaders, others are designers, others work more with the technical aspects. Deliberate negotiations of power are not currently part of the practice in Fab Lab Oulu.

The next questions are related to decision-making: **Who makes decisions? Why? Do all participants understand the reasons for decisions?** In short-term activities, almost everything is decided by our instructors. Pupils have certain creative freedom, but the processes are much guided. In long-term projects, children make most decisions. Fab lab instructors just define the goal of the activity (the expected outcomes), present the problems children must solve, and sometimes list the processes children may use. After that, children are free to work on their own project, choosing the methods that they desire with the support of instructors that would assist children when they need some advice. Hence, the vast majority of decisions are made by the pupils themselves.

Regarding reflection on the process and outcome, the following questions are asked: **Are there occasions for all participants for critical reflection on the process and the outcomes? Are there occasions for evaluation for all participants, on both individual and group level?** We consider reflection as an important part of learning. Because of that instructors in Fab Lab Oulu reserve time for public reflection where everybody is encouraged to provide their own thoughts after each working session: what tasks pupils have completed, what kind of challenges they faced, where they succeeded and where they failed. Sometimes teachers prepare a survey on the impact of the activity in the pupils and discuss the results later in their classroom. The reflection sessions make sense in long-term activities where children have had time to develop their ideas, to fail, and to find solutions to problems. These sessions do not seem to be easy for the children as they are not generally familiar with this kind of practice, and hence it is sometimes difficult to explain the value of reflection. Some children do not find the activity useful while others are very shy to talk. Hence, reflection sessions can be very difficult to conduct successfully and sometimes almost nothing comes out of it. Exploring different methodologies to make reflection sessions more successful would be beneficial, e.g., documenting step by step what was done in the activity.

9.4.5 Conditions for Competence

The conditions for competence emphasize increasing children's competence during the activities. The first questions asked, **what kind of responsibility children have/do not have? Why?** In Fab Lab Oulu activities, the main responsibility of children is to attend the activity and to perform the tasks. In long-term activities, children's responsibility is to operate the machines correctly after an introduction to the machine use and safety aspects. Instructors always follow the process though. We think this is part of the learning outcomes of the activity.

The next questions asked, **who defines the goals for the activity? Are children allowed to take part in defining the goals? Why/why not? Do all participants understand the goals? Does everybody get a chance to contribute? Do all participants listen to each other?** As has been already discussed, in short-term activities, the goal is set by the instructors. In long-term activities, the goal is usually to solve a problem. In that case, the goals are set by the instructor and/or teacher. Sometimes, pupils are asked to explore certain processes. In that case, they can define the goals of the activity themselves. When working in groups, children can decide how to achieve these goals, and sometimes they define subtasks with related goals and assign those to one or more members of the group. In these cases, children discuss the feasibility of the idea with instructors. We believe that as fab lab is a new environment for children it would be very challenging for them to define the goals of each activity. Some children come to fab lab to complete their own projects generally outside the school hours. In that case, children define themselves as what they want to do and how. Instructors will provide support when needed. To achieve this, children must have adequate skills in some digital fabrication processes. The scope of the activities is planned in such a way that all children can participate if they want to. Some children are more active than others are, but generally, all of them contribute somehow.

The next questions are related to information: **How much information do you provide in the beginning of the activities? Do children have all information they need? How can they get it?** At the beginning of the activities, Fab Lab Oulu instructors provide information on the goal of the activity and its physical outcome and functionality (if any); a short introduction to digital fabrication processes to be used, including the machines to use; examples of conducted projects; how to search for additional information (tutorials, keywords to use); schedule of the activity. During the activities, instructors give tips for information searching or show online tutorials.

The next question regarding conditions for competence asked, **do children's activities have real impact? Does the project result in tangible outcomes? Do children learn something? Does this learning build on top of previous knowledge/competences?** Children's activities really do have an impact. Something tangible is produced in all activities as that is one of the main goals of digital fabrication; this product is something that can be shown to others and can hopefully be a source of pride for children. In some cases, Fab Lab Oulu instructors showcase children's work in the fab lab 'display window' for everybody to see. Children have a possibility to give back something to the rest of the community as well, if they end up working at fab lab on their own time and helping then other visitors. In addition, some of the activities organized by teachers in Fab Lab Oulu involved the development of business plans and selling out the resulting products. Other schools have used funding obtained from selling products or organizing learning activities to buy new machines for their school makerspaces. Learning activities in fab lab can also be designed so that the goal is to build something that would help others or solve some existing challenges in the community.

At fab lab, children develop a new set of skills; they do not learn only technical and design skills but also some horizontal competencies such as creativity, critical thinking, problem-solving, computational thinking, and collaboration. The longer the activity the more it is possible to learn. Defining the learning outcome should be the first thing instructors should consider when defining an activity. When organizing workshops at fab lab, sometimes this goal is not addressed explicitly in the setup phase, and this can lead to a total failure. Learning at fab lab is built on top of previous knowledge although sometimes children are not aware of that. For instance, when they are doing a 2D design, they need to understand the metric system, the different geometric figures and what is a segment or a vertex. One big challenge for fab lab instructors is that they usually are not familiar with the background of children. That is why we believe that activities should be built in collaboration with teachers. Learning and competence development should be teacher-led as they have pedagogical competence. We see that they should also be the ones who evaluate children's learning.

The final question addresses an essential issue, **does the work process support children to initiate future projects by themselves?** Getting started with digital fabrication is perhaps the most difficult part. Getting to know the necessary software, machines and processes usually take some time. When children are familiar with the basic information, know where to find additional information, and know also the possibilities of the different processes it is easier for them to work on their own. We aim to help children to build confidence and feel empowered to try new things. Our instructors always remind children that they are welcome to come to fab lab whenever they want, and that fab lab staff is there to help them. We also explain that it is impossible that they know about all the topics, but the knowledge can be built by doing and by getting inspiration from others—standing on the shoulders of giants.

9.4.6 Summary of the Insights

In Table 9.2, we summarize the insight gained from our collaborative working, reflecting on the Fab Lab Oulu current practices in relation to the framework on the empowerment of children (Chawla and Heft 2002; Kinnula and Iivari 2019), in the form of aspects that we propose all fab labs to follow when preparing activities with schoolchildren.

9.5 Discussion

In this study, we wanted to understand the potential of fab labs in empowering children to make and shape digital technology, and what kind of best practices, limitations, or challenges can be identified.

9.5.1 Research Implications

The contribution of this study comes through a detailed, practice-based contemplation on the potential of fab lab as a site for empowering children to make and shape digital technology. Even if the studies have already brought up fab lab as a site for engaging children in the design and making activities (Blikstein and Krannich 2013; Iivari and Kinnula 2018; Iivari et al. 2018a; Iversen et al. 2016; Katterfeldt et al. 2015; Posch and Fitzpatrick 2012; Posch et al. 2010; Pucci and Mulder 2015), the particularities of fab labs as such a site have not been scrutinized. Particularly novel and valuable is the inclusion of fab lab personnel in the contemplation of these issues. They have years of practical experience working with children and teachers in design and making projects in the fab lab. Such an experience is now combined with a research-based understanding of the conditions for the empowerment of children

Table 9.2 Insights from the reflection of Fab Lab Oulu practices with the framework of empowerment of children

Conditions of convergence	Conditions of entry
Is it easy and natural for children to participate?	**Have the participants been fairly selected? Has somebody been excluded? Why?**
When organizing activities for schools, teachers should be involved in the activity. They know better the children, their strengths and weaknesses	Local government should be involved in advertising and organization of educational activities targeted to schools. They should facilitate the transporting of children to the premises. In that way, there is no discrimination due to the location of school or other social and economic factors
Do not replicate same processes as in the school. Give children a chance to work on their own and make their own decisions	When advertising activities it is important to reach a large number of schools as possible. Social networks and direct email messages to principals have worked in Fab Lab Oulu
Do not only organize activities for schools. Out of school activities might encourage children to continue to work on their own projects	Before hosting an activity, participants should be asked if they need some special arrangements. In that way, people with disabilities can also participate in the activity. Fab lab space should be as accessible as possible
Are existing organizations and structures supporting children's participation relied on?	**Is children's participation voluntary? Why? If not, why?**
It is important to build a community with local schools	When teachers include any fab lab activity in teaching, the activity is generally mandatory for pupils. When designing the activity, the instructors should consider that some participants are possibly not going to enjoy the activity and they should be prepared for that. Traditionally, activities run in fab labs are targeted at making enthusiasts and generally motivated people. This change might disturb some instructors
It is important to educate teachers on digital fabrication. They need to know fab lab potential to collaborate more actively in designing activities	
Educational activities can be outsourced to other organizations. They can reach such groups that are otherwise difficult to reach	
Do not focus on educational events only, it is important that the general public learns to know fab lab potential	
Use social media to promote events and encourage participants to use their own social networks to share their work	**Is the location and schedule for the activities easy to access for children and their families?**
Are activities based on children's own issues and interests?	When organizing a school activity in fab lab, collaborate with partners nearby to offer other activities as well (e.g., visiting research centers in a university)
In short-term activities, try to fabricate a personalized object, providing a clear layout	When organizing activities for school-age children, it is important that a fab lab is open during school hours as well as after school
In long-term activities, give children an appropriate level of freedom to choose what to fabricate	

(continued)

Table 9.2 (continued)

Conditions for competence	
What kind of responsibility children have/do not have? Why? Make children's responsibilities visible and link them to learning goals Give children the chance to use the machines after explaining them safety issues. It helps them to feel ownership of the activity	**Do children's activities have a real impact? Does the project result in tangible outcomes? Do children learn something? Does this learning build on top of previous knowledge/competences?** Consider such topics for projects that are meaningful, sustainable and address existing problems in the community Aim *always* for a personal tangible object as an outcome of activities Encourage children to publish and advertise documentation of the work on social media/websites Have an exhibition space in your fab lab to showcase children's work Plan the activities children's learning in mind and build it on top of their previous knowledge. Discuss with teachers children's background and what knowledge they already have
Who defines the goals for the activity? Are children allowed to take part in defining the goals? Why/why not? Do all participants understand the goals? Does everybody get a chance to contribute? Do all participants listen to each other? Let children define the goal of the project when possible. Be sure that the goals match with the learning outcomes specified at the beginning of the activity Prepare activities with enough complexity so everybody has something to do	**Does the work process to support children to initiate future projects by themselves?** Tell children that first they need to learn and understand the basics and then it is easier to see the potential of digital fabrication Help children to see digital fabrication as a possible hobby Make sure that fab lab is accessible to children outside of schoolwork
How much information do you provide at the beginning of the activities? Do children have all information they need? How can they get it? Specify children clearly the goal of the activity, expected learning outcomes, schedule and activity structure. In addition, it is recommended to give a short introduction to digital fabrication and the main processes children are using Try not to answer children's questions directly; teach them information searching processes instead and give pointers where to find more information. Follow their progress to assist when they are stuck	

(continued)

Table 9.2 (continued)

Conditions of social support	Conditions for reflection
Is the environment supportive? Is there a team spirit? How can this be encouraged? Do children support and encourage each other? How can this be supported? Are everybody's opinions and thoughts considered valuable? Activities at fab lab promote collaboration among children naturally. Blurring the role of teachers and instructors, converting them into facilitators, seems to foster collaboration among children A reflective discussion after a session might help to show children that their inputs are valued by the group. However, it is difficult to make all children participate actively in reflection **Are all participants respected? Do all participants act friendly and politely? How can this be encouraged?** In general, there is cordial and collaborative behavior among the participants. Teachers can help with behavioral problems, as they know the children	**Do power differences exist between participants? Have the power differences been deliberately negotiated?** Power differences in school and fab lab are different; the role of teachers and instructors becomes more blurred than in the school environment **Who makes decisions? Why? Do all participants understand the reasons for decisions?** In long-term activities, pupils usually make decisions; they decide how to tackle the proposed problem. In short-term activities, the instructors guide the pupils and make most decisions **Are there occasions for all participants for critical reflection on the process and the outcomes? Are there occasions for evaluation for all participants, on both individual and group level?** Reflection sessions are useful and should be run after each session. However, they are difficult to conduct

to make and shape digital technology. Through this, we managed to generate a rich and empirically grounded set of insights on fab lab best practices, limitations, and challenges around the empowerment of children. This enables taking the framework proposed by Kinnula and Iivari (2019) a step further. Kinnula and Iivari (2019) critically considered the conditions for the empowerment of children and proposed a set of questions to ask when aiming at empowering children to make and shape digital technology. This study offered a needed, practice-based evaluation and refinement of the questions. The resulting insights should be useful for practitioners working in fab labs as well as in other informal learning settings with children. In addition, the insights should be useful broadly for researchers interested in the empowerment of children to make and shape digital technology through design and making (e.g., Iversen et al. 2017).

9.5.2 Implications for Practice

Based on the insights presented in Table 9.2, we formulated a set of guidelines for practitioners arranging school visits to a fab lab or working with children in fab labs or more broadly in different kinds of non-formal learning settings.

When you are **preparing your fab lab for working with children**:

- Determine what are your goals and motivation for the work as they affect what you do and how you work with children.
- If teachers are involved, organize training also for them. It is important they understand the potential of digital fabrication.
- Organize every now and then events to the general public. It helps to advertise the fab lab, and that children are interested in attending to the activity. Use social media to promote events.
- Sometimes, organizing activities might be outsourced to other partners, for instance, startups.
- Sometimes schools or other organizations working with children face difficulties to gather resources for traveling to fab lab. Try to involve local administration and organizations, in such a way they can put adequate resources, and avoid discrimination due to location or economic factors.
- Consider your resources, including space to work in, personnel, machine time, physical materials, needed knowledge, and education. Be sure that they are adequate enough.

When you are **planning the activities**:

- Remember to ask if there are children that need any special arrangement.
- When possible involve teachers in activity preparation. It is always better if teachers have some training in digital fabrication. It is important that they understand the potential of digital fabrication and how it can be integrated into the

curriculum. Discuss with teachers what is the background of the children. Define learning outcomes, methodologies and goals of the activity together.

- Learning outcomes and goals should be thought carefully. This should provide the basis to build the activity on. Consider that children are going to learn from each other at the same time. Consider also as learning outcomes horizontal competencies such as creativity, critical thinking, collaboration, or computational thinking.
- Be sure that at the end of an activity children build something tangible, better if they can take it home. Aim for the object to be something purposeful: something that can be used later by children themselves or that could help others.
- Use appropriate teaching methods: project-based learning and learning by doing work better for long-term activities; tutorial-type activity works better if time is shorter.
- Activities with schools are usually mandatory for all children. When preparing such activities consider that not all children are going to enjoy it. Try to build a contingency plan for this situation.
- In short-term activities, it is better if the goals and methods are defined by instructors. In longer term activities, children should try to define the goals and methods by themselves. Instructors could take part in the decisions, guiding children according to what is feasible or not.
- Consider the number of members in workgroups. Adapt the goals according to that.

Consider the following **during an activity**:

- At the beginning of the activity describe very clearly to children (and teachers): Expected learning outcome; Goal of the activity—what physical object you expect to have after the activity; Short introduction to the digital fabrication process/processes to be used including how to use the machines; Duration and structure of the activity, including the reflection sessions; How children can find more information: which online resources are available, which keywords they could use to find information.
- Inform children about how to operate the machines and safety instructions. Afterward, when possible, let children operate the machines. Operating machines safely is an important responsibility for them.
- Advertise other activities that children can do outside school hours (if any).
- Encourage participants to use social media to share their work.
- Understanding the roles of instructors and teachers is important. There is an important change to what children are familiar with: Initially, an instructor is seen as a teacher, but when children understand the mechanics they are seen as facilitators; Teachers, if present, can assist instructors. They know better the strengths and weaknesses of the children. From the children's perspective, teachers have more authority than instructors do, so teachers can help to correct some misbehavior. Anyhow, the power difference between teacher and children is reduced in fab lab environment.

- The instructor should act as a facilitator, not as a classical teacher. Provide pointers so children can find the answers, do not answer their questions directly. Let children make mistakes.
- Conduct a group reflection after each session, trying that all children participate. This should make them realize the things they have learnt and that their ideas could help others. Note that organizing a good reflection session is challenging. Children are not familiar with them, and some children are very shy to talk. Asking children to document their work might help to conduct the reflection session.
- Teachers, if present, can help to form balanced working groups. Inside a group, we prefer to let children assign roles by themselves. Be flexible, roles might change during an activity.

After an activity consider these:

- Try to run a questionnaire both to teachers and children: what worked, what did not work. This will help to improve the activity.
- When possible, make public all content related to the activity, with pictures of the resulting objects and even children's/teacher's opinions. This will help other fab labs that are in the same situation.
- Try to have the fab lab open outside school hours. This will help children with special interest to make any kind of project at fab lab on their own time.

9.6 Conclusions and Future Directions

In this paper, we have provided rich practical insight and guidelines for working at fab lab with schoolchildren and in collaboration with schools. Behind these guidelines is our firm belief that learning to design and make digital technologies is empowering for children as such and that with careful consideration of the working practices it is possible to further support children's empowerment and help them to make and shape their technology-rich world. We hope that these guidelines are helpful for both fab lab personnel—instructors and managers alike—as well as teachers or city administrative staff who plan to work in collaboration with a local fab lab.

This study is limited by its focus on practices of a single fab lab in Finland. Fab Lab Oulu has had extensive collaboration with local schools, however. For future research, we suggest examining further the roles of all different stakeholders somehow related to fab labs and how they can, on their part, help in making the collaboration between schools and fab labs a seamless whole with equal opportunities for all schools and children, regardless of where they reside.

Acknowledgements We express our thanks to Antti Mäntyniemi for his contribution to this work. This research has been funded by European Union's Horizon 2020 Research and Innovation programme under grant agreement No. 787476 (COMnPLAY SCIENCE project) as well as by Academy of Finland project funding under grant agreement No. 324685 (Make-A-Difference). It is also connected to the GenZ project, a strategic profiling project in human sciences at the University of Oulu. The project is supported by the Academy of Finland (Grant #318930) and the University of Oulu.

References

Ananiadou, K., & Claro, M. (2009) 21st century skills and competences for new millennium learners in OECD countries. In *OECD Education Working Papers* (Vol. 41). Paris: OECD.

Bar-El, D., & Zuckerman, O. (2016). *Maketec: A makerspace as a third place for children*. Paper presented at the The 10th International Conference on Tangible, Embedded, and Embodied Interaction, Eindhoven, Netherlands.

Blikstein, P. (2013). Digital fabrication and 'making' in education: The democratization of invention. In J. Walter-Herrmann & C. Büching (Eds.), *FabLab: Of machines, makers and inventors* (pp. 1–21). Transcript Verlag.

Blikstein, P., & Krannich, D. (2013). *The makers' movement and FabLabs in education: Experiences, technologies, and research*. Paper presented at the International Conference on Interaction Design and Children (IDC'13), New York, NY, USA.

Chawla, L., & Heft, H. (2002). Children's competence and the ecology of communities: A functional approach to the evaluation of participation. *Journal of Environmental Psychology, 22*(1–2), 201–216.

Chu, S. L., Quek, F., Bhangaonkar, S., Ging, A. B., & Sridharamurthy, K. (2015). Making the maker: A means-to-an-ends approach to nurturing the maker mindset in elementary-aged children. *International Journal of Child-Computer Interaction, 5*(Sept), 11–19.

Dougherty, D. (2016). *Free to make: How the maker movement is changing our schools, our jobs, and our minds*. North Atlantic Books.

Druin, A. (2002). The role of children in the design of new technology. *Behaviour and Information Technology, 21*(1), 1–25.

Druin, A., Bederson, B., Boltman, A., Miura, A., Knotts-Callahan, D., & Platt, M. (1999). Children as our technology design partners. In A. Druin (Ed.), *The design of children's technology* (pp. 51–72). San Francisco: Kaufmann.

Eshach, H. (2007). Bridging in-school and out-of-school learning: Formal, non-formal, and infromal education. *Journal of Science Education and Technology, 16*(2), 171–190.

Gershenfeld, N. (2012). How to make almost anything: The digital fabrication revolution. *Foreign Affairs* (91), 43–57.

Gershenfeld, N., Gershenfeld, A., & Cutcher-Gershenfeld, J. (2017). *Designing reality: How to survive and thrive in the third digital revolution*. Basic Books.

Iivari, N., & Kinnula, M. (2016). *Inclusive or inflexible—a critical analysis of the school context in supporting children's genuine participation*. Paper presented at the Nordic Conference on Human-Computer Interaction (NordiCHI'16), Gothenburg, Sweden.

Iivari, N., & Kinnula, M. (2018). *Empowering children through design and making—towards protagonist role adoption*. Paper presented at the International Conference on Participatory Design (PDC'18), Hasselt and Genk, Belgium.

Iivari, N., Kinnula, M., & Molin-Juustila, T. (2018a). *You have to start somewhere—initial meanings making in a design and making project*. Paper presented at the International Conference on Interaction Design and Children (IDC'18), Trondheim, Norway.

Iivari, N., Kinnula, M., Molin-Juustila, T., & Kuure, L. (2018b). Exclusions in social inclusion projects: Struggles in involving children in digital technology development. *Information Systems Journal, 28*(6), 1020–1048.

Iivari, N., Molin-Juustila, T., & Kinnula, M. (2016). *The future digital innovators: Empowering the young generation with digital fabrication and making*. Paper presented at the International Conference on Information Systems (ICIS'16), Dublin, Ireland.

Iversen, O. S., & Smith, R. (2012). *Scandinavian participatory design: Dialogic curation with teenagers.* Paper presented at the International Conference on Interaction Design and Children (IDC'12), Bremen, Germany.

Iversen, O. S., Halskov, K., & Leong, T. W. (2010). *Rekindling values in participatory design.* Paper presented at the 11th Biennial Participatory Design Conference (PDC'10), Sydney, Australia.

Iversen, O. S., Smith, R. C., Blikstein, P., Katterfeldt, E. S., & Read, J. C. (2016). Digital fabrication in education: Expanding the research towards design and reflective practices. *International Journal of Child-Computer Interaction, 5,* 1–2.

Iversen, O. S., Smith, R. C., & Dindler, C. (2017). *Child as protagonist: Expanding the role of children in participatory design.* Paper presented at the International Conference on Interaction Design and Children (IDC'17), Stanford, California, USA.

Iversen, O. S., Smith, R. C., & Dindler, C. (2018). *From computational thinking to computational empowerment: A 21st century PD agenda.* Paper presented at the Participatory Design Conference (PDC), Hasselt and Genk, Belgium.

Katterfeldt, E. S., Dittert, N., & Schelhowe, H. (2015). Designing digital fabrication learning environments for Bildung: Implications from ten years of physical computing workshops. *International Journal of Child-Computer Interaction, 5,* 3–10.

Kinnula, M., & Iivari, N. (2019). *Empowered to make a change: Guidelines for empowering the young generation in and through digital technology design.* Paper presented at the FabLearn Europe conference, Oulu, Finland.

Kinnula, M., Iivari, N., Isomursu, M., & Kinnula, H. (2018). Socializers, achievers or both? Roles of children in technology design projects. *International Journal of Child-Computer Interaction, 17*(Sept), 39–49.

Kinnula, M., Iivari, N., Molin-Juustila, T., Keskitalo, E., Leinonen, T., Mansikkamäki, E., Käkelä, T., Similä, M. (2017). *Cooperation, combat, or competence building—what do we mean when we are 'empowering children' in and through digital technology design?* Paper presented at the International Conference on Information Systems (ICIS'17), Seoul, South-Korea.

Kinnula, M., Laari-Salmela, S., & Iivari, N. (2015). *Mundane or magical? Discourses of technology adoption in Finnish schools.* Paper presented at the European Conference on Information Systems (ECIS'15), Münster, Germany.

Laru, J., Vuopala, E., Iwata, M., Pitkänen, K., Sanchez, I., Mäntymäki, A., et al. (2019). Designing seamless learning activities for school visitors in the context of Fab Lab Oulu. In Looi CK., Wong LH., Glahn C., Cai S. (eds) *Seamless Learning* (pp. 153–169). Lecture Notes in Educational Technology. Singapore: Springer.

Litts, B. K. (2015). *Resources, facilitation, and partnerships: Three design considerations for youth makerspaces.* Paper presented at the International Conference on Interaction Design and Children conference (IDC'15), Boston, Massachusetts, USA.

NCBE. (2014). *Finnish national core curriculum for basic education* (in Finnish). Retrieved from https://www.oph.fi/download/163777_perusopetuksen_opetussuunnitelman_per usteet_2014.pdf

OECD. (2012). *Connected minds: Technology and today's learners.* Retrieved from https://read. oecd-ilibrary.org/education/connected-minds_9789264111011-en

Pitkänen, K., Iwata, M., & Laru, J. (2019). *Supporting Fab Lab facilitators to develop pedagogical practices to improve learning in digital fabrication activities.* Paper presented at the FabLearn Europe Conference, Oulu, Finland.

Posch, I., & Fitzpatrick, G. (2012). *First steps in the FabLab: Experiences engaging children.* Paper presented at the Australian Computer-Human Interaction Conference (OzCHI'12).

Posch, I., Ogawa, H., Lindinger, C., Haring, R., & Hörtner, H. (2010). *Introducing the FabLab as interactive exhibition space.* Paper presented at the International Conference on Interaction Design and Children (IDC'10), Barcelona, Spain.

Pucci, E. L., & Mulder, I. (2015). Star(t) to Shine: Unlocking Hidden Talents Through Sharing and Making. *Distributed, Ambient, and Pervasive Interactions,* 85–96.

Read, J. C., Fitton, D., & Horton, M. (2014). *Giving ideas an equal chance: Inclusion and representation in participatory design with children.* Paper presented at the International Conference on Interaction Design and Children (IDC'14), Aarhus, Denmark.

Read, J. C., Horton, M., Fitton, D., & Sim, G. (2017). *Empowered and informed: Participation of children in HCI.* Paper presented at the Human-Computer Interaction (INTERACT 2017)—16th IFIP TC 13 International Conference, Mumbai, India.

Read, J. C., Horton, M., Sim, G., Gregory, P., Fitton, D., & Cassidy, B. (2013). *CHECk: A tool to inform and encourage ethical practice in participatory design with children.* Extended Abstracts on Human Factors in Computing Systems (CHI'13), Paris, France.

Read, J. C., & Markopoulos, P. (2013). Child–computer interaction. *International Journal of Child-Computer Interaction, 1*(1), 2–6.

Sánchez Milara, I., Pitkänen, K., Niva, A., Iwata, M., Laru, J., & Riekki, J. (2019). *The STEAM path: Building a community of practice for local schools around STEAM and digital fabrication.* Paper presented at the FabLearn Europe Conference, Oulu, Finland

Smith, R. C., Iversen, O. S., & Hjorth, M. (2015). Design thinking for digital fabrication in education. *International Journal of Child-Computer Interaction, 5*(September), 20–28.

Spreitzer, G. M. (1995). An empirical test of a comprehensive model of intrapersonal empowerment in the workplace. *American Journal of Community Psychology, 23*(5), 601–629.

Thomas, K. W., & Velthouse, B. A. (1990). Cognitive elements of empowerment: An "interpretive" model of intrinsic task motivation. *Academy of Management Review, 15*(4), 666–681.

Van Mechelen, M., Sim, G., Zaman, B., Gregory, P., Slegers, K., & Horton, M. (2014). *Applying the CHECk tool to participatory design sessions with children.* Paper presented at the International Conference on Interaction Design and Children (IDC'14), Aarhus, Denmark.

Warschauer, M. (2002). Reconceptualizing the digital divide. *First monday, 7*(7).

Weibert, A., & Schubert, K. (2010). *How the social structure of intercultural computer clubs fosters interactive storytelling.* Paper presented at the 9th International Conference on Interaction Design and Children (IDC'10).

Ylioja, J., Georgiev, G. V., Sánchez, I., & Riekki, J. (2019). *Academic recognition of fab academy.* Paper presented at the FabLearn Europe Conference, Oulu, Finland

Marianne Kinnula is an Associate Professor at the University of Oulu and the vice-leader of the INTERACT Research Unit. Her research is in the fields of Information Systems and Human–Computer Interaction; she is interested in how technology changes our everyday lives in many ways, at society level, organizational level, as well as at the individual level. School as an organization and children's technology use are close to her heart and, currently, the main focus of her studies: what is children's genuine possibility to affect the decisions that concern them as well as their technology-rich environment. She holds an editorial position in the International Journal of Child–Computer Interaction and has published actively in leading HCI and IS conferences and journals.

Netta Iivari is a Professor in Information Systems and research unit leader of the INTERACT Research Unit at the University of Oulu. She has a background in Cultural Anthropology as well as in Information Systems and Human–Computer Interaction. Her long-lasting research interest concerns understanding and strengthening people's participation in shaping and making their digital futures. Her research has addressed packaged and open source software development contexts as well as the empowerment of children through design and making. Her research is strongly influenced by interpretive and critical research traditions. She has a specific interest in the development and utilization of culture and discourse-oriented lenses as well as in the examination and support of transdisciplinary research and design. She has received research project funding from numerous funding sources such as the EU and the Academy of Finland. She holds editorial positions in respected HCI and IS journals and conferences and she has actively published in high-quality IS and HCI journals and conferences.

Iván Sánchez Milara is a Ph.D. student in the Center for Ubiquitous Computing, University of Oulu. At the same time, he is working as a fab lab instructor at Fab Lab Oulu. Sánchez Milara's research is focused on the interaction between teachers and technology. He is studying what kind of digital tools teachers need to be able to generate their own interactive and tangible learning content. He has authored more than 40 peer-reviewed research articles in international venues. He is also involved in building a local community of teachers interested in digital fabrication with the goal of integrating these processes in formal education. In addition, he teaches different university courses related to digital fabrication.

Jani Ylioja is the founder and director of Fab Lab Oulu; he is an expert in community digital fabrication. His work is connected globally to fab labs including Fab Foundation and MIT Center of Bits and Atoms, the original source of fab labs. He is the acting chairperson for Nordic Fab Labs Network Association. He acts as a technical expert in several projects including Futurena (to bioprint kidneys) and Novidam (Novel materials for digital fabrication for electronics, optics, and medical solutions). He lectures several courses on digital fabrication at the University of Oulu.

Part V
Synopsis and Research Agenda

The final part includes a chapter providing a synopsis of the book and research agenda for the future development in the field.

Chapter 10
Science Learning in the ICT Era: Toward an Ecosystem Model and Research Agenda

Michail N. Giannakos

Abstract This chapter closes this edited volume on *Non-Formal and Informal Science Learning in the 21st Century*. Through an ecosystem perspective, we aim to understand and represent the interrelationships among the ecosystem elements that provide actors with avenues by which they may be introduced to and become knowledgeable about science and science learning. This is particularly relevant for non-formal and informal learning contexts since actors engage in science learning activities outside the formal learning context, and therefore they are not (necessarily) learning and teaching professionals, and also science education is not (necessarily) their main objective (e.g., when in informal learning contexts). In addition, actors are different from one another; therefore, it is necessary to take into consideration their attributes and beliefs to better understand their behavior, their capabilities, and their needs, which in turn will improve efficiency, coherence, and performance of the ecosystem overall. The overarching goal of this chapter is to present a conceptualization of informal and non-formal science education through an ecosystem model and propose a research agenda for the future. By doing this, the chapter seeks to offer a broader foundation for paving the way toward a holistic understanding of *Non-Formal and Informal Science Learning in the 21st Century*.

Keywords Informal learning · Non-formal learning · Science education · Ecosystem model

10.1 Introduction

The term "ecosystem" has been introduced to describe a system that includes living organisms, their non-living environment, and all their interrelationships in a particular unit of space (Tansley 1935). The term has been applied to different fields such as biology, technology, and education. The concept of the ecosystem (i.e., interactions

M. N. Giannakos (✉)
Norwegian University of Science and Technology (NTNU), Trondheim, Norway
e-mail: michailg@ntnu.no

© Springer Nature Singapore Pte Ltd. 2020   181
M. Giannakos (ed.), *Non-Formal and Informal Science Learning
in the ICT Era*, Lecture Notes in Educational Technology,
https://doi.org/10.1007/978-981-15-6747-6_10

between organisms and their environment) has the capacity to employ representations that may be used in education research to conceptualize educational systems (Giannakos et al. 2016; Barron 2006; National Research Council 2014; Traphagen and Traill 2014). The environment (context) may be physical or not, and includes activities, material resources, relationships, and the interactions that emerge from them (Barron 2006). The concept of an ecosystem in the context of learning puts the learner at the center of the system and allows us to focus on activities and relationships across settings and time (Bell et al. 2009). The conceptualization of science (or science, technology, engineering, and mathematics; STEM) learning in the form of an ecosystem is not new (Traphagen and Traill 2014; Corin et al. 2017), but it is arguable that it provides both the language to discuss an inclusive learner-centered system and the roadmap to develop collaborations between organizations and groups in the future (Corin et al. 2017).

Science learning can be seen as an ecosystem where the actors actively interact and collaborate with each other to create knowledge and new capacities while evolving their interrelations, leading to novel pedagogical frameworks and technological affordances. The advances in information and communications technology (ICT) as well as the inter- and multidisciplinary nature of science education offer diverse opportunities for non-formal and informal science learning. A comprehensive understanding of the science education ecosystem and its interdependencies will allow us to identify potential barriers as well as enable us to develop frameworks and technological affordances that will provide solutions that benefit the different actors within the ecosystem.

10.2 The Potential of the Science Education Ecosystem Approach in Non-Formal and Informal Settings

As has been described in the literature, besides the main actors (organisms: teachers, parents, etc.), a science learning ecosystem might also include various organizations (e.g., schools, science centers, civil society; see, e.g., Traphagen and Traill 2014; Corin et al. 2017). Through the learning ecosystem perspective, we aim to understand and represent the interrelationships among the ecosystem elements that provide actors with avenues by which they may be introduced to and become knowledgeable about science (Corin et al. 2017). This is particularly relevant for non-formal and informal learning contexts since actors engage in science activities outside the formal learning context and the knowledge obtained is transferred and enriched between contexts (Barron 2006; Traphagen and Traill 2014). Another important element that posits the ecosystem perspective as a sound metaphor to describe non-formal and informal science education is the fact that it adopts the "porous" nature (Traphagen and Traill 2014) of the boundaries between learning settings (compared to the relatively siloed nature of formal learning settings).

The representation of science education as an ecosystem highlights that each actor/organization complements and builds upon each other's efforts (Traphagen and Traill 2014). Such a system working at full capacity has been envisioned to distribute responsibility for teaching and learning among all of the ecosystem's elements (National Research Council 2014). To sustain collaborations over time, science learning ecosystems must be attentive to what Traphagen and Traill (2014) term the "enlightened self-interest" of their members; participating in the ecosystem must allow members to work toward their own organization's goals, objectives, and missions. Therefore, the alignment and co-existence of self-interests are critical and allow the various actors, as well as the ecosystem overall, to reach their own goals efficiently. In this chapter, we use the concept of the ecosystem to understand and represent the interrelationships among the various organisms (actors and organizations), the enablers, and the development of particular attitudes, values, and dispositions that young people as learners and as citizens may develop, in the context of informal and non-formal science education.

10.3 Conceptualizing Science Education and Its Ecosystem in Non-Formal and Informal Settings

As already mentioned, in this work we adopt a perspective that recognizes the interconnectedness of an "ecosystem" and the aspects of learner agency within such a complex system, focusing on the ecosystem of science education. Previous works identify patterns of exclusion in science education, including contemporary forms of stereotypes, sexism, and other modes of inequality (e.g., Lord et al. 2019; Master et al. 2016). An educational ecosystem can be described as a set of complex self-organized communities that consist of actors that have different attributes, decision principles, and beliefs (Tsujimoto et al. 2018). Furthermore, an ecosystem consists of multiple hierarchical layers, cooperation, and collaboration; in addition, competition among its different actors is found to be of great importance, but difficult to achieve (Pappas et al. 2018). The relations among the different actors and organizations of an educational ecosystem cannot remain solely within the learning and teaching context; instead, they are likely to extend to different contexts, like personal, business, or procedural relations. Since the actors and organizations involved are different from one another, it is important to explore their attributes to better understand their behavior, expectations, capabilities, and needs, which in turn can be orchestrated to improve the efficiency and coherence of the ecosystem overall.

When referring to education and learning, the term ecosystem describes the environment created and supported by the numerous actors and organizations that comprise the ecosystem, as well as their interactions and interrelations. Gibson (1986) demonstrates how the understanding of the environment empowers potentialities for action (e.g., doors are openable). That work highlights the functional significance (affordances or enablers) that is visible to individuals (actors) with reciprocal skills (effectivities) and the intention to act (Gibson 1986). While the environment provides such potentialities, their meaning can only be materialized through actor–environment interaction. Therefore, being an affordance or enabler is a property of an ecosystem. In other words, "The environment is a closed (but unbounded) set of affordances, or functionally defined goals, that identify the potential perceptions of the animal [individual actor] and that complement the effectivities" (Turvey and Shaw 1979, p. 206). Educational ecosystems inherit the concept of a learning ecology; that is, "the set of contexts found in physical or virtual spaces that provide opportunities for learning. Each context is comprised of a unique configuration of activities, material resources, relationships and the interactions that emerge from them" (Barron 2006, p. 195). Since our goal is to create sustainable ecosystems that promote science learning, we need to take into account the various actors and organizations, their capabilities, goals, and needs, as well as the potentialities of the environment.

Interactions among the various actors (e.g., teachers, policymakers) and the environment (e.g., government) are essential to creating the needed technological, institutional, and pedagogical conditions. Building on the above discussion, we posit that a science education ecosystem model comprises organisms, which can be individual actors (e.g., children, parents, instructors, curators) or organizations (e.g., schools, museums, universities, industry), who all have capabilities, goals, and needs. The actors and organizations need to utilize the various enablers that are available in their respective contexts, which will not only lead to the development and alternation of actors' motivations, beliefs, and self-efficacy but also affect the society and business development. This is an iterative process based on which the organisms use available enablers to constantly achieve their goals, and in our case to promote science learning. Figure 10.1 presents the Science Education Ecosystem (SEE) model, which conceptualizes the organisms that need to cooperate, coordinate, and collaborate through the utilization and orchestration of the various enablers (e.g., means and activities), focusing on the potential for nurturing scientifically informed behaviors and improving attitudes, values, and dispositions that young people possess about science and science education.

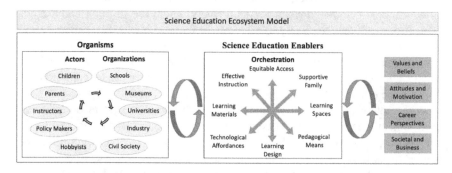

Fig. 10.1 The science education ecosystem (SEE) model

10.4   Conclusions and the Way Ahead

The importance of novel enablers such as digitalization and the utilization of emerging learning spaces is going to attract a lot of attention in science learning in the upcoming years. Novel technologies and spaces will enable and democratize science learning practices that can empower different actors (e.g., instructor, parent, hobbyist) to contribute to the ecosystem of science education. The proposed SEE model is an attempt to conceptualize these interrelationships and provide actors with avenues for facilitating learning and societal change, therefore generating knowledge that impacts both contemporary science learning practices and the society overall.

In this closing part of the volume, we would like to highlight two research avenues that are critical for the future development of non-formal and informal science learning in the twenty-first century.

The role of actors and organizations in utilizing and further developing science learning practices. How actively may the various actors (e.g., teachers, parents, policymakers) and organizations (e.g., schools, companies, universities) be involved in order to shape the development of novel science learning practices? These actors and most organizations are typically involved in bottom-up self-interested endeavors, through which they are introducing novel learning spaces and arenas, as well as a set of evidence-based practices that have been optimized through continuous planning, implementation, evaluation, and refinement. Therefore, the actors involved are furthering contemporary practices that benefit the ecosystem, as well as the particular contexts and learning settings (formal, non-formal, and informal) and science content areas (e.g., problem-solving, manufacturing, coding).

Adoption and integration of new practices and affordances. Future research needs to examine how different actors (e.g., teachers, policymakers) and organizations (e.g., science centers, schools) can be empowered to adopt and integrate novel practices and affordances in their established processes. This is critical for the ecosystem to be able to utilize new knowledge. For such adoption and integration to succeed, various measures, such as personnel training and renewal of routines, need to be implemented.

Acknowledgements This work is supported by the "Learning science the fun and creative way: coding, making, and play as vehicles for informal science learning in the 21st century" Project, under the European Commission's Horizon 2020 SwafS-11-2017 Program (Project Number: 787476) and the "Learn to Machine Learn" (LearnML) project, under the Erasmus+ Strategic Partnership program (Project Number: 2019-1-MT01-KA201-051220).

References

Barron, B. (2006). Interest and self-sustained learning as catalysts of development: A learning ecology perspective. *Human Development, 49*(4), 193–224. https://doi.org/10.1159/000094368

Bell, P., Lewenstein, B., Shouse, A., & Feder, M. (2009). *Learning Science in Informal Environments: People, Places, and Pursuits.* Committee on Learning Science in Informal Environments. Washington, DC: National Academies Press.

Corin, E. N., Jones, M. G., Andre, T., Childers, G. M., & Stevens, V. (2017). Science hobbyists: Active users of the science-learning ecosystem. *International Journal of Science Education, Part B, 7*(2), 161–180.

Giannakos, M. N., Krogstie, J., & Aalberg, T. (2016). Video-based learning ecosystem to support active learning: Application to an introductory computer science course. *Smart Learning Environments, 3*(1), 11.

Gibson, J. J. (1986). *The Ecological Approach to Visual Perception.* Hillsdale, NJ: Lawrence Erlbaum.

Lord, S. M., Ohland, M. W., Layton, R. A., & Camacho, M. M. (2019). Beyond pipeline and pathways: Ecosystem metrics. *Journal of Engineering Education, 108*(1), 32–56.

Master, A., Cheryan, S., & Meltzoff, A. N. (2016). Computing whether she belongs: Stereotypes undermine girls' interest and sense of belonging in computer science. *Journal of Educational Psychology, 108*(3), 424.

National Research Council. (2014). *STEM Learning Is Everywhere: Summary of a Convocation on Building Learning Systems.* Washington, DC: National Academies Press.

Pappas, I. O., Mikalef, P., Giannakos, M. N., Krogstie, J., & Lekakos, G. (2018). Big data and business analytics ecosystems: Paving the way towards digital transformation and sustainable societies. *Information Systems and e-Business Management, 16*, 479–491.

Tansley, A. G. (1935). The use and abuse of vegetational concepts and terms. *Ecology, 16*(3), 284–307.

Traphagen, K., & Traill, S. (2014). *How Cross-Sector Collaborations Are Advancing STEM Learning.* Los Altos, CA: Noyce Foundation.

Tsujimoto, M., Kajikawa, Y., Tomita, J., & Matsumoto, Y. (2018). A review of the ecosystem concept—Towards coherent ecosystem design. *Technological Forecasting and Social Change, 136*, 49–58.

Turvey, M. T., & Shaw, R. E. (1979). The primacy of perceiving: An ecological reformulation of perception for understanding memory. In L. G. Nilsson (Ed.), *Perspectives on Memory Research* (pp. 167–222). Hillsdale, NJ: Lawrence Erlbaum.

Michail N. Giannakos is a Professor of interaction design and learning technologies at the Department of Computer Science of NTNU, and Head of the Learner-Computer Interaction lab (https://lci.idi.ntnu.no/). His research focuses on the design and study of emerging technologies in online and hybrid education settings, and their connections to student and instructor experiences and practices. Giannakos has co-authored more than 150 manuscripts published in peer-reviewed journals and conferences (including Computers & Education, Computers in Human Behavior, IEEE TLT, Behaviour & Information Technology, BJET, ACM TOCE, CSCL, Interact, C&C, IDC to mention few) and has served as an evaluator for the EC and the US-NSF. He has served/serves in various organization committees (e.g., general chair, associate chair), program committees as well as editor and guest editor on highly recognized journals (e.g., BJET, Computers in Human Behavior, IEEE TOE, IEEE TLT, ACM TOCE). He has worked at several research projects funded by diverse sources like the EC, Microsoft Research, The Research Council of Norway (RCN), US-NSF, German agency for international academic cooperation (DAAD) and Cheng Endowment; Giannakos is also a recipient of a Marie Curie/ERCIM fellowship, the Norwegian Young Research Talent award and he is one of the outstanding academic fellows of NTNU (2017–2021).